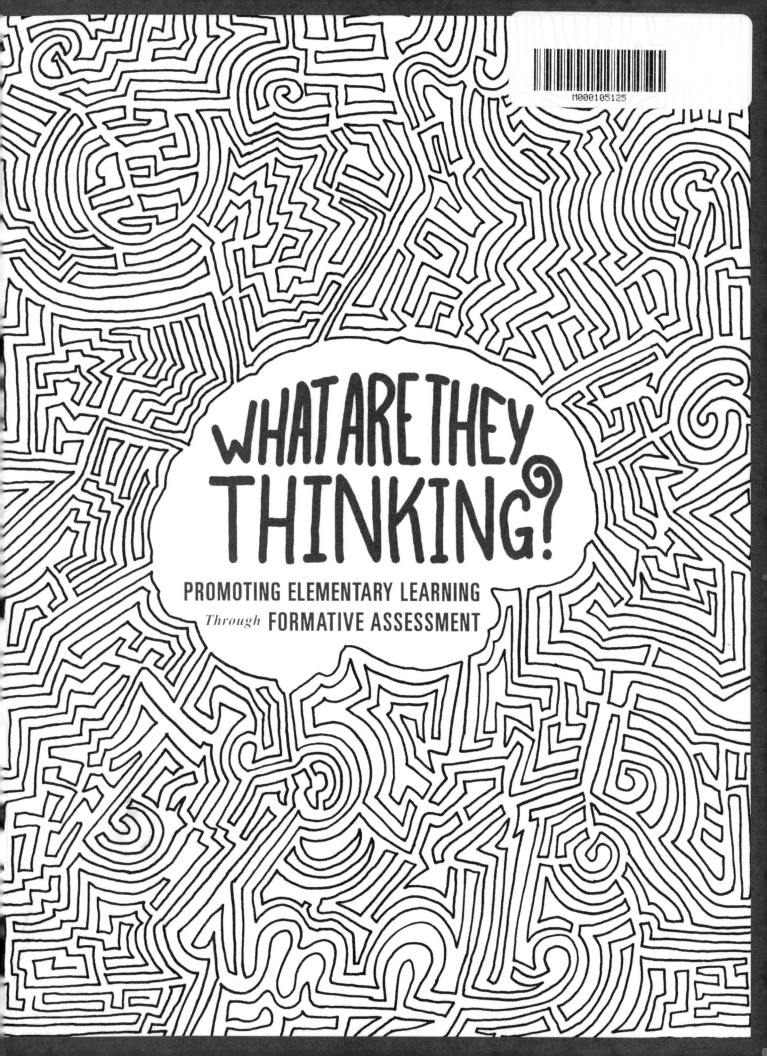

WHAT ARE THEY THINKING?

PROMOTING ELEMENTARY LEARNING
Through FORMATIVE ASSESSMENT

WHAT ARE THEY THINKING?

PROMOTING ELEMENTARY LEARNING
Through FORMATIVE ASSESSMENT

PAGE KEELEY

National Science Teachers Association

Arlington, Virginia

National Science Teachers Association

Claire Reinburg, Director
Wendy Rubin, Managing Editor
Andrew Cooke, Senior Editor
Amanda O'Brien, Associate Editor
Amy America, Book Acquisitions Coordinator

ART AND DESIGN
Will Thomas Jr., Director
Rashad Muhammad, Graphic Designer, cover and interior design

PRINTING AND PRODUCTION
Catherine Lorrain, Director

NATIONAL SCIENCE TEACHERS ASSOCIATION
David L. Evans, Executive Director
David Beacom, Publisher

1840 Wilson Blvd., Arlington, VA 22201
www.nsta.org/store
For customer service inquiries, please call 800-277-5300.

NSTA is committed to publishing material that promotes the best in inquiry-based science education. However, conditions of actual use may vary, and the safety procedures and practices described in this book are intended to serve only as a guide. Additional precautionary measures may be required. NSTA and the authors do not warrant or represent that the procedures and practices in this book meet any safety code or standard of federal, state, or local regulations. NSTA and the authors disclaim any liability for personal injury or damage to property arising out of or relating to the use of this book, including any of the recommendations, instructions, or materials contained therein.

Library of Congress Cataloging-in-Publication Data
Keeley, Page.
  What are they thinking? : promoting elementary learning through formative assessment / by Page Keeley.
    pages cm
  Includes bibliographical references and index.
    ISBN 978-1-938946-25-7 -- ISBN 978-1-938946-65-3 (electronic) 1. Education, Elementary--United States. 2. Education, Elementary--United States--Evaluation. 3. Education, Elementary--Aims and objectives. 4. Educational evaluation--United States.. 5. Inquiry-based learning--United States. I. Title.
  LA219.K44 2014
  372--dc23
                    2014007864

Contents

Contents

About the Author

PAGE KEELEY recently retired from the Maine Mathematics and Science Alliance (MMSA) where she was the senior science program director for 16 years, directing projects and developing resources in the areas of leadership, professional development, linking standards and research on learning, formative assessment, and mentoring and coaching. She has been the principal investigator and project director of three National Science Foundation grants: the Northern New England Co-Mentoring Network, PRISMS: Phenomena and Representations for Instruction of Science in Middle School; and Curriculum Topic Study: A Systematic Approach to Utilizing National Standards and Cognitive Research. She has also directed state Math-Science Partnership projects including TIES K–12: Teachers Integrating Engineering into Science K–12 and a National Semi-Conductor Foundation grant, Linking Science, Inquiry, and Language Literacy (L-SILL). She also founded and directed the Maine Governor's Academy for Science and Mathematics Education Leadership, a replication of the National Academy for Science and Mathematics Education Leadership of which she is a Cohort 1 Fellow.

Page is the author of 16 national bestselling books, including four books in the *Curriculum Topic Study* Series, nine volumes in the *Uncovering Student Ideas in Science* series, and three books in the *Science Formative Assessment: 75 Practical Strategies for Linking Assessment, Instruction, and Learning* series. Currently she provides consulting services to school districts and organizations throughout the United States on building teachers' and school districts' capacity to use diagnostic and formative assessment. She is a frequent invited speaker on formative assessment and teaching for conceptual change.

Page taught middle and high school science for 15 years before leaving the classroom in 1996. At that time she was an active teacher leader at the state and national level. She served two terms as president of the Maine Science Teachers Association and was a District II NSTA Director. She received the Presidential Award for Excellence in Secondary

Science Teaching in 1992, the Milken National Distinguished Educator Award in 1993, the AT&T Maine Governor's Fellow in 1994, the National Staff Development Council's (now Learning Forward) Susan Loucks-Horsley Award for Leadership in Science and Mathematics Professional Development in 2009, and the National Science Education Leadership Association's Outstanding Leadership in Science Education Award in 2013. She has served as an adjunct instructor at the University of Maine, was a science literacy leader for the AAAS/Project 2061 Professional Development Program, and serves on several national advisory boards. She is a science education delegation leader for the People to People Citizen Ambassador Professional Programs, leading the South Africa trip in 2009, China in 2010, and India in 2011.

Prior to teaching, she was a research assistant in immunology at the Jackson Laboratory of Mammalian Genetics in Bar Harbor, Maine. She received her BS in life sciences from the University of New Hampshire and her MEd in secondary science education from the University of Maine. In 2008 Page was elected the 63rd president of the National Science Teachers Association (NSTA).

Foreword

Page Keeley's popular book series "Uncovering Student Ideas..." led to the creation of a column in *Science and Children* offering probes and strategies for use by elementary teachers. This column has been popular since its inception with readers eager to use each probe in their classrooms. The popularity is due to several characteristics of Page's probes—they are based on research, aligned with grade band expectations, easily implemented, and effective.

Probes uncover students' ideas. Every child brings prior understanding to each learning experience, whether incomplete, incorrect, or fully developed. It is critical to understand what children bring to the learning environment both before proceeding with instruction and during lessons to provide scaffolding and regulate rigor. Every step of the way, throughout the development of conceptual understanding, we must have a clear picture of what students have learned and the understanding they are developing. This is the vital role of Page's assessment probes.

Never has formative assessment been more important than now. With the release of the *Next Generation Science Standards* and shifts in state standards there is new emphasis on progressions of learning and conceptual understanding. *A Framework for K–12 Science Education: Practices, Crosscutting Concepts, and Core Ideas* is specific: "To support students' meaningful learning in science and engineering, all three dimensions need to be integrated into standards, curriculum, instruction, and assessment" (NRC 2012, p. 2). It goes on to mention the role of formative assessment explicitly and points out the need for this component in teacher professional development to enhance the capacity of schools to implement effective science curriculum.

Once you have read and used the suggestions provided in any of the chapters in this collection, you will find many applications that support your understanding of student knowledge while also providing new learning opportunities. The variety of formats; the manner in which they are embedded in instructional models; and the strategies they provide to create opportunities for science talk, initiating investigations, eliciting questions, self assessment, and reflection on learning are all applicable to many settings.

Page has selected a variety of probes to share with you that provide a wide perspective. Beginning with how to use the book she goes on to provide ways in which professional learning, by both individuals and communities, can effectively be created. Once you have studied and used some of Page's suggestions, I know you will join me in thanking her for sharing this wonderful collection of strategies and materials. *What Are*

They Thinking? is specifically designed to support elementary teachers as they assess student understanding and Page has provided the insight and knowledge to assure you of success in supporting your students moving toward conceptual understanding.

Linda Froschauer
Editor, *Science and Children*
NSTA President, 2006–2007

REFERENCE

National Research Council (NRC). 2012. *A framework for K–12 science education: Practices, crosscutting concepts, and core ideas*. Washington, DC: National Academies Press.

Introduction

The Role of Formative Assessment in Elementary Science: An Overview

Formative assessment in science is a process that informs instruction and supports learning through a variety of tools and techniques aimed at uncovering and examining students' thinking, then using that information to drive instruction that supports students moving toward conceptual understanding of the learning target. Formative assessment in science is inextricably linked to learning. As teachers are collecting information about students' thinking related to core concepts and phenomena, students are simultaneously constructing new understandings, revising prior beliefs, and strengthening their ability to engage in the practices of science and engineering. Formative assessment is frequently referred to as assessment *for* learning rather than assessment *of* learning. One can even add a third preposition—assessment *as* learning. As you will see in each of these chapters, the formative assessment tool or technique highlighted is in essence a learning activity for students, as well as a way for the teacher to gather information about students' ideas and ways of thinking in order to determine next steps in moving their learning forward.

Elementary science teachers face an added challenge using formative assessment to move students' learning forward. Because science is a way of understanding our natural world, students arrive with ideas that have already been formulated based on their everyday experiences outside the classroom. These experiences begin in infancy, long before students enter the formal classroom. Daily experiences and interactions with "felt weight," moving objects, shapes, light and shadows, observations of living things, dropping objects, "disappearing" materials, visible changes, and other phenomena are already shaping children's ideas at an early age. Therefore, one important use of formative assessment in science is uncovering the preconceptions students bring with them to their learning, as these preconceptions will often affect the way students think about new information. Children make sense of the content they encounter in the science classroom in their own way, based on their interactions with the natural world, the words they encounter in their daily conversations and in various media, the materials they use, and the contexts in which they learn.

By taking the time to understand students' thinking at any point during an instructional cycle, instruction becomes more focused and deliberate in moving students toward an intended learning target. It begins by helping students think about—and then verbalize or write about—their existing ideas, giving them and the teacher a springboard from which to launch into instructional experiences that will build a bridge between where students are and where they need to be. Targeted instruction then confronts students' existing ideas, providing opportunities for them to test their ideas through investigation and engage in productive science talk that incorporates scientific reasoning, construction of scientific explanations, and argumentation supported by evidence. As the teacher uses formative assessment to monitor changes in students' ideas and ways of thinking, students often resolve the conflict between their initial ideas and new ways of thinking. This process is called *conceptual change* and is strongly supported by the use of formative assessment throughout a full cycle of instruction. It begins with the elicitation of students' initial ideas and ends with reflection on new knowledge and changes in thinking.

However, to use formative assessment to promote learning in the elementary classroom, the teacher must have access to a repertoire of formative assessment classroom techniques (FACTs) and specially designed questions that link research on learning to core concepts in science (probes). But having access to these tools at your fingertips is not enough. Teachers also need to understand *how* these tools are appropriately used for formative assessment and what formative assessment looks like in the elementary classroom. That is the purpose of this book—to build and support elementary teachers' capacity to use formative assessment tools to link assessment, instruction, and learning in the science classroom.

This book will help elementary teachers deepen their understanding of their students' thinking in order to promote conceptual learning in the K–6 classroom. It moves instruction away from the pervasive practice of selecting an activity first to instead starting with an understanding of students' ideas and then selecting an appropriate activity to begin instruction. It helps teachers make adjustments to their instructional materials throughout the cycle of instruction. This is very different from following "the script" without understanding if the "script" is the right match to where students are in their thinking. The focus is on what the student is thinking and learning, not the materials or activities. This is the difference between teaching science as a series of "hands-on" activities and teaching science for conceptual understanding.

Elementary teachers are the first line of offense in addressing common misconceptions that follow students from elementary grades into middle school, into high school, and even into adulthood. This is why it is so important to build elementary teachers' capacity to continuously and seamlessly use formative assessment in science. If elementary students are provided with opportunities to resolve the inconsistencies between their way of thinking and the scientific way of thinking, many of the difficulties that students encounter in later grades as they progress through increasingly complex ideas and ways of thinking can be eliminated. Clearly, this is why elementary teachers are important to developing science-literate high school graduates, well prepared and interested in entering STEM fields in college or in the workplace. Elementary science teachers are critical links in a K–12 system of science learning. This book is intended to support you in that critical role!

Organization of This Book

This book is organized into 30 chapters. Each chapter features an article written for the NSTA *Science and Children* journal's monthly column, "Formative Assessment Probes: Promoting Learning Through Assessment." Each article features a formative assessment probe from one of the eight books in the NSTA Press series *Uncovering Student Ideas in Science*.

A probe is a two-tiered assessment specifically designed to reveal common misconceptions. It begins with an engaging prompt situated in a familiar context, followed by a set of selected responses. Many of the distracters in the selected responses mirror the research on children's alternative conceptions. The number of distracters used depends on the number of research-identified misconceptions. The probes avoid the use of technical terminology in order to uncover students' conceptual understanding and not their memorization of definitions. The selected response is then followed by a section in which students explain their thinking by constructing an explanation. It is this part of the probe that reveals the reasoning students use to make sense of a concept or phenomenon. It also provides insight into how a student's misconception may have developed: from their experiences in and out of the classroom, the words they encounter, their intuition, the context in which previous learning took place, or from their misinterpretation during the teaching and learning process.

Probes are often combined with a FACT (formative assessment classroom technique). FACTs are used in a variety of formats, ranging from individual formative assessment to uncovering student ideas within a small group or during a whole-class discussion. FACTs and probes are embedded throughout an instructional cycle of engagement and elicitation, exploration of ideas, formal concept development, application, and reflection. They fit easily within a 5E model of instruction or any variety of instructional models that use a learning cycle approach. FACTs serve a variety of teaching and learning purposes, including engaging and motivating students, eliciting preconceptions, activating thinking and metacognition, providing stimuli for productive science talk, initiating investigations, determining learning transfer, improving the quality of questions and responses, providing feedback, peer and self-assessment, and post-assessment or reflection on learning.

The articles in each chapter were specifically written to illustrate how a formative assessment probe, often combined with a FACT, is used in a K–6 classroom. While each of the books in the *Uncovering Student Ideas in Science* series provides K–12 teacher notes that accompany each probe, the teacher notes in the book series do not provide extensive descriptions of how the probe is used in an elementary classroom, actual examples of student work or transcripts of students talking about their ideas, or illustrative examples of instructional decisions made by elementary teachers. The article included in each chapter provides this information specifically for elementary teachers, giving deeper insight into the formative assessment process and complementing the teacher notes. In addition, a link is provided at the end of the chapter that will take the reader to a website where they can download a copy of the probe to use with their students (Note: Only the probe is provided in each of the links in the chapters. The teacher notes for each probe are found in the referenced book in which the probe was originally published).

Each chapter also includes a Reflection and Study Guide. These guides include a set of questions designed to help the reader reflect on what they learned after reading the chapter. The questions can also be used for pre-reading. This is followed by a section on "Putting Formative Assessment Into Practice" that can be used after teachers try out the probe with their students. This section guides teachers in examining their own students' thinking and reflecting on their use of the formative assessment probe or FACT. A final set of questions in the "Going Further" section can be used to extend professional learning by suggesting other resources for individual or collaborative group learning. Many of these suggestions include links to *A Framework for K–12 Science Education* or the *Next Generation Science Standards*. Even if your state has not adopted the *NGSS*, the links provided will help you clarify the content in your own standards and provide you with a lens to focus on what effective teaching and learning in science involves when using the formative assessment probes.

Suggestions for Using This Book

The primary purpose of this book is to improve and support the teaching and learning of elementary science by embedding formative assessment into daily instruction. This purpose can be met through an individual teacher's use of this book or through collaborative structures for teacher learning. The following are suggestions for ways to use this book as a teacher, teacher leader, mentor, science specialist, professional development provider, or preservice instructor.

Overall Use of This Book

- Use Table 1 (pp. xvi–xvii) to match your own instructional or professional learning objective to the focus of the chapter and the probe that is highlighted. Read the chapter and use the Reflection and Study Guide questions to deepen your learning and inform your instruction before you teach a curricular unit.

- If you have not used the probe before, answer the probe yourself before reading the chapter. By experiencing the process of thinking through your own ideas, you may better understand what your students experience as they think through their ideas.

- If you have access to the *Uncovering Student Ideas in Science* books, read the teacher notes after completing the chapter. The teacher notes provide additional details on curricular considerations, related research, connections to national standards, and instructional suggestions that complement the chapter.

- After reading a chapter and using the guiding questions, note what you will do differently in your classroom as a result. Also note any information or suggestions to share with colleagues at the school or district level.

- If you are a classroom teacher or have access to students, try out the probe or FACT with children and compare what you experienced and learned through your own students with the chapter description or classroom vignette.

- Use the "Going Further" suggestions to extend your learning after reading the chapter. Search the internet or the NSTA Learning Center (*http://learningcenter.nsta. org*) for additional resources to continue your learning related to the chapter.

Structures for Professional Learning

- Chapters can be used within a workshop format to address content or a teaching strategy. Select chapters that match the professional learning goal of a workshop.

- Professional Learning Communities (PLCs) can select chapters for reading, discussion, and application to their professional goals as a PLC.

- Form study groups to learn about, try out, examine, and improve upon techniques for formative assessment. Choose a chapter to read, discuss, try out, and report back on its use.

- Create a classroom video of your use of a probe or FACT discussed in one of the chapters. Share and discuss the chapter with peers. Use the video to discuss and provide constructive feedback on the use of the probe or FACT.

TABLE 1. FEATURED PROBES AND TOPICS

Chapter	Probe	Topic
1	Doing Science	Scientific investigation; examine how misuse of the "scientific method" impacts students' ideas about the nature of science
2	Floating Logs	Floating and Sinking; use of intuitive rules to reason about floating and sinking
3	Does It Have a Life Cycle?	Life cycles; addressing the limitations of context in the curriculum
4	What Is a Hypothesis?	Hypotheses; revealing misconceptions teachers have about the nature of science that can be passed on to students
5	How Far Did It Go?	Linear measurement; difficulties students have with measurement, particularly with a non-zero starting point
6	Needs of Seeds	Seed germination and needs of living things; engaging in evidence-based argumentation
7	The Mitten Problem	Energy transfer, heat, insulators; teaching for conceptual change and how children's everyday experience affects their thinking
8	Is It Living?	Characteristics of living things; examine ways to uncover "hidden meanings" students have for some words and concepts in science
9	Various probes	A variety of probes and FACTs are used to show purposeful links to various stages in an assessment, instruction, and learning cycle
10	Emmy's Moon and Stars	Solar system, relative distances; importance of examining students' explanations even when they choose the right answer; impact representations have on children's thinking
11	Talking About Forces	Forces; examining common preconceptions and use of language to describe forces and motion
12	Is It an Animal?	Biological conception of an animal; explore how formative assessment probes can be used to engage in teacher action research
13	Pond Life	Single-celled organisms; use of representations to examine students' ideas
14	Objects in the Sky	Seeing the Moon in daytime; Challenges the adage "seeing is believing" with "believing is seeing"—examines reasons why children hold on to their strongly held beliefs
15	Can It Reflect Light?	Light reflection; addressing students' preconceptions with firsthand experiences that support conceptual change

Chapter	Probe	Topic
16	Is It Food for Plants?	Food, photosynthesis, needs of plants; using bridging concepts to address gaps in learning goals, understanding students' common sense ideas
17	Where Did the Water Come From?	Condensation; Using the water cycle to show how a probe can be used to link a core content idea, scientific practice, and a cross-cutting concept.
18	Catching a Cold	Infectious disease, personal health; Using a probe to uncover common myths and folklore related to the common cold
19	Me and My Shadow	Sun-Earth System; using a formative assessment probe to engage students in productive science talk
20	Birthday Candles	Light transmission, connection between light and vision; using drawings to support explanations
21	Mountain Age	Weathering and erosion; organizing student data using a classroom profile for instructional decisions and professional development
22	Solids and Holes	Floating and sinking; density; using the P-E-O technique to launch into investigations
23	Chrysalis	Life cycle; over-emphasis on labeling diagrams with correct terminology may mask conceptual misunderstandings related to the life cycle of a butterfly.
24	Batteries, Bulbs, and Wires	Electrical circuits; lighting a bulb with a battery and a wire; how science kit materials may make it difficult for students to examine the way a complete circuit works
25	Is It a Rock?	Rocks; human-made materials; how continuous assessment is used throughout an instructional unit.
26	Is It a Solid?	Solids and liquids; using the claim card strategy to uncover ideas and support students' construction of explanations using claims and evidence.
27	When Is the Next Full Moon?	Moon cycle, using the concept cartoon format to uncover ideas before making observations
28	Swinging Pendulum	Motion patterns; using formative assessment of a scientific idea to assess readiness for an engineering problem
29	Is It Melting?	Melting and dissolving; illustrating use of a probe in professional development to uncover misconceptions and make formative decisions for teacher learning
30	Is It Made of Parts?	Structure of organisms; scaffolding formative assessment of a learning target by identifying and assessing sub-ideas

- Combine a chapter with the use of a protocol for looking at student work (LASW). The chapter can provide the groundwork or information for next steps after teachers discuss student work.

- Create a seminar series that features a chapter for each session. Use Socratic dialogue to discuss the chapter.

- Teachers who mentor new teachers can read and discuss chapters together. The new teacher can try out the probe or FACT and reflect on his or her learning with the mentor teacher. The mentor teacher can provide useful feedback as a link between the chapter and the novice teachers' practice.

- Lesson study groups can use the chapters to inform the design of the lesson they will use. Discussions during the debriefing of the lesson can be linked back to the chapter.

- Form a book study group, face-to-face or electronically. Select chapters for the book study. Use the Reflection and Study Guide for in-person or online discussions.

- Share and discuss a chapter during a grade-level team or faculty meeting. Discuss how the example of formative assessment could be applied to other disciplines.

- Curriculum planning committees can use the chapters to consider ways to embed formative assessment into the elementary curriculum. Use the chapter to provide implementation support for the curriculum.

- Conduct collaborative action research with a colleague. Choose a chapter and design a classroom research project related to the chapter. Use the example and suggestions in Chapter 12 (Teachers as Classroom Researchers) to engage in teacher research using the probes.

- Use a formative assessment probe for teacher learning in workshops or other settings. After teachers respond to the probe, use ideas from the chapter to make formative decisions or engage teachers in practices such as argumentation.

- Select chapters that can support teachers in implementing the *NGSS* or their state standards. Use the chapter for discussions about formative assessment and learning targets.

- Consider writing your own article about the use of a formative assessment probe or FACT. Use the examples in this book to help you. Consider sharing your article with your school colleagues or submitting for publication in NSTA's *Science and Children*.

- Use the book as a whole-group jigsaw book study during teacher institutes. Assign chapters to pairs or small groups as reading assignments. Each group can prepare a short presentation to teach what they learned to others.

- Preservice instructors can use this book as a required text in their courses or select specific chapters to integrate into their courses.

- Science curriculum coordinators can use chapters to support teachers working toward improving their practice.

- The *Science Formative Assessment* book series (Keeley 2008, 2014) has been frequently used in a variety of professional learning formats. Select chapters that highlight use of a FACT and combine with the reading from *Science Formative Assessment*.

- Select a probe that can be used across grade levels. Administer the probe, collect data on students' thinking, and engage colleagues in cross-grade-level data-driven discussion. Use the chapter to ground the group's discussion about the probe and students' ideas.

- Come up with your own idea for ways to use these chapters for professional learning that builds teachers' understanding of how to use formative assessment effectively.

Teacher Learning Outcomes

Whichever chapters you decide to use in this book or the variety of ways you decide to use them, consider the following outcomes:

1. You may learn new content about the science you teach. Everyone has misconceptions, including teachers (*all* teachers, not just elementary teachers). These chapters might surface long-held misconceptions you were not aware that you had. Working through and resolving these misconceptions is a significant part of your professional learning.

2. You will learn a lot about your own students. Although the chapters describe scenarios of students in other classrooms, it is quite likely your students will think and respond in a similar way. Furthermore, trying out the probes with your own students will give you insight into your own students' thinking and how similar their ideas are to what has been learned from research about children's ideas in science.

3. You will learn new instructional strategies that link assessment and instruction, which will help you build a rich repertoire of effective teaching practices.

4. You will increase your capacity to implement the disciplinary core ideas, the scientific and engineering practices, and the crosscutting concepts in the *NGSS*. Even if your state has not adopted the *NGSS*, your teaching and your students' learning connected to your state's standards will be enhanced through your knowledge of formative assessment practices connected to the *NGSS*.

5. You will bring new vitality and food for thought to collaborative teacher learning by sharing and discussing these chapters and the use of the probes with your colleagues in a variety of professional learning formats. The value in sharing the probes, your student data, your inquiries into practice, and your new learning with other teachers cannot be overstated. It is transformative and will lead to significant changes in practice among teachers at all levels of experience, within and across classrooms.

Continuing Your Learning

When you finish the book, your learning about using formative assessment probes in the elementary classroom does not end with the final chapter. Continue to check out Page Keeley's formative assessment probe column each month in *Science and Children*, as well as articles written by other authors that feature the use of a formative assessment probe or formative assessment strategies. Visit the *Uncovering Student Ideas in Science* website at *www.uncoveringstudentideas.org* for additional ideas. Go to the NSTA Learning Center (*http://learningcenter.nsta.org*) and participate in discussions with other teachers who are using the probes and formative assessment strategies. Attend an NSTA area or national conference and look for sessions on formative assessment, including sessions presented by the author of this book and her colleagues. And finally, share the ways you have used this book by contacting the author, Page Keeley, at *pagekeeley@gmail.com*. By sharing your ideas, together we can build a collaborative community to promote elementary science learning through formative assessment.

Chapter 1
"Doing" Science

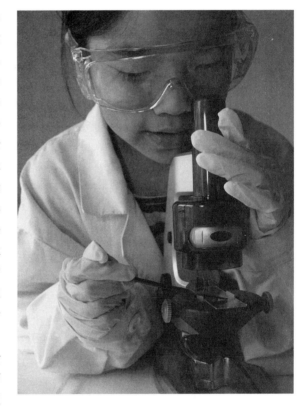

From the very first day of school, children should be involved in not only using science to investigate the world around them but also learning how scientists practice science. How many of you begin the school year by introducing your students to the various ways scientists engage in their work? Perhaps you begin with what is often the first topic in your textbook, "The Scientific Method." When planning lessons that address the nature of science, have you ever taken the time to find out what students really think about how scientists conduct investigations? Could students' previous experiences have led to strongly held erroneous beliefs about how science is practiced?

The "Doing Science" probe from *Uncovering Student Ideas in Science, Volume 3: Another 25 Formative Assessment Probes* (Keeley, Eberle, and Dorsey 2008) can reveal some surprising ideas your students have about how scientists do their work. In order to build conceptual understanding that leads to a deep appreciation of the way science is practiced, you must start by uncovering the preconceived ideas your students bring to the science classroom. The "Doing Science" probe (Figure 1.1, p. 3) is designed to elicit commonly held ideas students have about the way scientists go about their investigations.

Many students believe there is a common series of steps that all scientists follow. According to William McComas (1996) "The notion of a single scientific method is so pervasive it seems certain that many students must be disappointed when they discover that scientists do not have a framed copy of the steps of the scientific method posted high above each laboratory workbench." Another commonly held student idea is that scientific investigations always involve doing "experiments." These ideas about the way scientists investigate can be quite tenacious and tend to follow students from one grade level to the next if left unchallenged.

About the Probe

The "Doing Science" probe (Figure 1.1) is an example of a *friendly talk* probe, in which students analyze others' thinking and choose the person with whom they most agree. They then provide a justification for why they agreed with one person and disagreed with the others. The probe is used to find out whether students recognize that scientists use different methods depending on their question (Marcos's response). Scientists' approach to investigating questions involves a well-thought-out, planned, methodical process, unlike Antoine's response. It is not a definite series of steps that all scientists follow (Tamara's response). Avery is partially right by saying that scientists use different methods, but she is incorrect in saying they all involve doing experiments.

Results from this probe are likely to reveal two common misunderstandings your students may have. The first is the belief that there is a rigid, step-by-step method, which is often conveyed as a result of teachers requiring students to follow "the scientific method" and writing a lab report that always uses a prescribed format. The second is the belief that scientists always do experiments. This may result from the overuse of the word *experiment*. Experiments are one type of investigation and usually involve testing cause–effect relationships between variables.

One way to adapt this probe is to make a bridge between "school science" and the science scientists practice. Ask students to select the response from the probe that best matches their classroom experiences in conducting investigations. This can be done as a paper–pencil task or as a class discussion. Ask them to describe examples from their science class investigations that match the answer they chose. The information can be quite insightful in revealing the disconnect between "school science" and the actual practice of science.

Make It Formative

To make the probe formative, you must next use the information to plan your next steps for instruction. If many students in your class chose Tamara, vary the ways students investigate using different methods by including opportunities to conduct field or remote observations, experiments, use models, and collect specimens. Repeatedly emphasize that different investigations use different methods, but they are all planned out and methodical. In addition, vary the way students record and write about their investigations. Using science notebooks is an authentic way to model how scientists

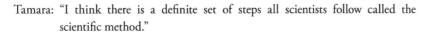

FIGURE 1.1.
The Doing Science Probe

Doing Science

Four students were having a discussion about how scientists do their work. This is what they said:

Antoine: "I think scientists just try out different things until something works."

Tamara: "I think there is a definite set of steps all scientists follow called the scientific method."

Marcos: "I think scientists use different methods depending on their question."

Avery: "I think scientists use different methods but they all involve doing experiments."

Which student do you most agree with? _____

Explain why you agree with that student and include why you disagree with the other students.

keep a record of their investigations. Also, be careful how you refer to "the scientific method." Simply changing *the* to *a* and referring to "a scientific method" implies that there is a method, but it is not the only one.

If several students chose Avery, that is an indication that you should be careful how you use the word *experiment*. Unless you are testing an idea, which involves a "fair test" for younger children and identifying and controlling variables for older elementary students, it is better to use the word *investigation*. Remember, all experiments are investigations, but not all investigations are experiments. Another way to help students is to expose them to different types of scientists and areas of science that are often observational and do not involve doing experiments such as astronomy, paleontology, and geology.

Breaking Misconceptions

Merely correcting students does not change their perception of science. They must have opportunities to recognize and experience how science is conducted. Unfortunately, many "cookbook" activities use a common procedural approach that helps reinforce the notion of a single scientific method. If you find your students have developed these misconceptions about how science is done, you should carefully look at the way science is portrayed both explicitly and implicitly in the textbooks and instructional materials you use. If you notice that your instructional materials may form or reinforce these common misconceptions, consider ways to turn these activities into more open-ended types of investigations in which students have to figure out a method to systematically investigate the phenomenon. Fortunately, many of the new instructional materials have discarded the traditional scientific method approach in favor of a broader depiction of the methods of science, including an emphasis on the importance of creativity in designing investigations.

The elementary science classroom is the first line of offense in making sure mis-representations of science do not shape students' views of the nature of science. Formative assessment tools help you become more aware of the ideas your students bring to the classroom or develop through the activities they experience or material they read in their textbooks. They may even reveal common misconceptions you have! When students' and your own thinking are made visible, appropriate decisions can be made to guide instruction that will help all students develop an accurate conception of the myriad ways scientists engage in their pursuit of new knowledge.

REFERENCES

Keeley, P., F. Eberle, and C. Dorsey. 2008. *Uncovering student Ideas in science, volume 3: Another 25 formative assessment probes.* Arlington, VA: NSTA Press.

McComas, W. 1996. Ten myths of science: Reexamining what we think we know. *School Science and Mathematics.* 96: 10.

INTERNET RESOURCE

Uncovering Student Ideas in Science series
www.nsta.org/publications/press/uncovering.aspx

NSTA CONNECTION

Read the introduction to *Uncovering Student Ideas in Science, Volume 1,* and download a full-size "Doing Science" probe at *www.nsta.org/SC1009.*

● ●

"Doing" Science
Reflection and Study Guide

QUESTIONS TO THINK ABOUT AFTER READING THIS CHAPTER

1. Think about how you would have answered this probe prior to reading this chapter. How did your K–12 science education influence your beliefs about the way science is practiced?

2. Why do you think students' notion of a single scientific method is so pervasive in K–12 education?

3. The probe provides four answer choices that mirror commonly held ideas students have about the way scientists investigate. Based on what you know about your students, what might be a fifth answer choice?

4. How do you think activities such as science fairs or lab reports with a specified format may contribute to the misconception that science always involves following the same series of steps?

5. What are some examples of investigations your students do in science class that are considered experiments? What are some examples of investigations your students do that are not considered experiments? What can you do to make sure your students understand that science involves a variety of different ways to investigate the natural world?

6. What do you think "school science" means? How is that similar or different from science as practiced by professional scientists?

7. How does referring to *a* scientific method differ from referring to *the* scientific method? Do you think changing the *to* a will make much of a difference in the way children think about science investigations?

8. The word *experiment* is an example of a word that has a different meaning in our everyday language versus the way it is used in science. Can you think of a way the word *experiment* is encountered by students in everyday conversations that is different from the way the word is used in science? Can you think of other examples of words used in our everyday language that differ from the way students encounter the same word in science?

9. Children's books that portray scientists and their work are excellent resources to help students understand the variety of ways scientists investigate the natural world. Which children's books have you used to portray scientists' work? How

could you use NSTA's annual list of Outstanding Science Tradebooks to help students learn about the nature of science. (See *www.nsta.org/publications/ostb*)

10. *A Framework for K–12 Science Education* (NRC 2012) and *The Next Generation Science Standards* (NGSS` Lead States 2013) emphasize a set of scientific and engineering practices rather than sets of procedures or process skills. How is this approach an improvement over the way "doing science" has traditionally been taught? (For more information about scientific and engineering practices, visit NSTA's *NGSS* portal at *http://ngss.nsta.org*)

PUTTING FORMATIVE ASSESSMENT INTO PRACTICE

1. What did you learn about your students' perception of "doing science" by examining their responses to the probe? Were you surprised by any of their responses?

2. Do any of their responses indicate where their ideas about "doing science" came from? How is knowing the origin of their "doing science" ideas helpful in making formative decisions about next steps for instruction?

3. What will you do next to address your students' misconceptions about "doing science"?

4. What modifications will you make to your curriculum, instruction, or instructional materials to ensure students develop an accurate understanding of how scientists investigate the natural world?

5. What can you do as a post-assessment to find out how your students' ideas about "doing science" have changed?

6. Based on what you learned from using the probe with your students, what suggestions do you have for your colleagues and future teachers?

GOING FURTHER

1. Read and discuss the Teacher Notes for the "Doing Science" probe (Keeley, Eberle, and Dorsey 2008, pp. 94–100). Pay particular attention to the Related Research and Suggestions for Instruction and Assessment sections.

2. Read and discuss the section on "Understanding How Scientists Work" on pages 43–44 in *A Framework for K–12 Science Education* (NRC 2012) or online at *www.nap. edu/openbook.php?record_id=13165&page=43.*

3. Read and discuss the article: Schwartz, R. 2007. What's in a word? How word choice can develop (mis)conceptions about nature of science. *Science Scope* 31 (2): 42–47.

4. Read and discuss William McComas's article Ten Myths of Science (McComas 1996). Article is reprinted at *www.amasci.com/miscon/myths10.html*

5. Read and discuss the *Science for All Americans* (AAAS 1988) description of scientific inquiry at *www.project2061.org/publications/sfaa/online/chap1.htm#inquiry*

6. Watch and discuss the NSTA archived web seminar on the practice of planning and carrying out scientific investigations *http://learningcenter.nsta.org/products/symposia_seminars/NGSS/webseminar7.aspx*

Chapter 2
"More A–More B" Rule

Floating and sinking provides opportunities for elementary students to explore and construct understandings about the properties of objects and materials before they build more sophisticated understandings of relative density and buoyancy at the middle school level. Floating and sinking phenomena are not new to elementary students. Before they are even introduced to the topic in school, students have often developed their own rules for determining whether an object or material floats or sinks. Some of these rules are incompatible with scientific explanations and are commonly referred to as *misconceptions*. Floating and sinking misconceptions often develop during early childhood experiences such as putting objects in water during play or while in the bathtub. The preconceptions students form from these early experiences strongly influence their thinking in the science classroom. As a result, it is important for elementary teachers to use formative assessments that reveal students' preconceptions and inform the teacher about next steps to take in their instruction.

By starting with students' preconceptions, revealed through the use of a formative assessment probe, teachers can scaffold inquiry-based experiences that will confront children with their misconceptions and guide them through a process of conceptual change. Carefully designed conceptual change instruction enables students to willingly give up their misconceptions in favor of new ideas that make sense to them. Typically students will not abandon their strongly held misconceptions until they have accepted a new idea that provides a better explanation than their old one. As long as a misconception makes sense to students, they will cling to it, even after instruction and well into adulthood.

One of the commonly held misconceptions students have about floating and sinking is the idea that large things sink and small things float. This misconception is hard to give up unless students have had opportunities to realize for themselves that this rule does not always hold true. This early-formed preconception can be explained by one of the intuitive rules identified by Stavy and Tirosh (2000) called *More A–More B*. Children encounter many everyday situations in which objects or materials are of different sizes and children have to determine the effect of size on the behavior or properties of that object or material. "More A–More B" is a basic, intuitive logical scheme children use to figure out these relationships. For example, children know if they have two rocks of the same kind, then the larger rock will weigh more. If they have a large container and a small container, then they know the larger container will hold more water. This rule works with extensive properties like mass, weight, and volume that depend on the amount of matter. However, it does not work with an intensive property such as density, which is the same for identical materials regardless of their size.

About the Probe

The "Floating Logs" probe (Figure 2.1) in *Uncovering Student Ideas in Science, Volume 2: 25 More Formative Assessment Probes* (Keeley, Eberle, and Tugel 2007) can be used in an elementary "Floating and Sinking" unit to find out whether students think objects made of the same material float differently when their size increases. This probe is based on a research study in which students ages 8–12 were asked how a longer candle would float compared to a shorter candle. Most students thought the longer candle would either sink or float lower than the smaller candle (Driver et al. 1994).

Before using the probe, show students a picture of a grape and a watermelon, preferably side by side. These materials are chosen because their shapes are the same but their size different (their material is also different, but students focus first on the observable shape and size). Ask students which one they think will sink and which one will float when they are put in water. Most students think the watermelon will sink and the grape will float. When asked why, a typical response is, "Watermelons are much bigger than grapes." When the two are put in water, students are quite surprised to see that the watermelon floats and the grape sinks. Before students develop the concept of "heavy for its size," they will often justify the observation by saying it is because the watermelon and grape are two different things made of different materials. If the materials are the same, then the larger one will sink. Use this student reasoning to lead into the "Floating Logs" probe (Figure 2.1).

The "Floating Logs" probe can be used with the predict-explain-observe (P-E-O) strategy. Students work in small groups to come to an agreed-upon prediction for how

FIGURE 2.1.
The Floating Logs Probe

Floating Logs

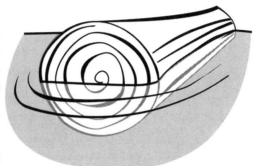

A log was cut from a tree and put in water. The log floated on its side so that half the log was above the water surface. Another log was cut from the same tree. This log was twice as long and twice as wide. How does the larger log float compared with the smaller log? Circle the best answer:

A More than half of the larger log floats above the water surface.

B Half of the larger log floats above the water surface.

C Less than half of the larger log floats above the water surface.

Explain your thinking. Describe the "rule" or the reasoning you used for your answer.

much the larger log will float compared to the smaller log. They then provide an explanation to support their prediction. Many students use a version of the "More A–More B" rule to make and justify the most common prediction: less than half of the log floats above the water surface. Typical explanations are "The log is bigger so it can't float as well in the water"; "The bigger log presses on the water more and makes it sink down further"; or "The bigger log needs more water to stay up."

Using materials that are similar to the log context in the probe, such as a small wooden dowel and a much thicker and longer wooden dowel, students can test their prediction and observe how much of the dowel floats above the water. However, don't be surprised if students will not believe their eyes! Some students will claim the dowel isn't big enough and ask for a larger piece. Others will say that wood floats the same but different sizes of other materials will not float the same. Still others might claim it depends whether the log is on its side or upright or that there needs to be more water for the larger piece of wooden dowel. Instead of correcting students, have them test their alternative ideas, if possible. Give them pairs of different solid materials of different sizes. Eventually students will see that solid objects made of the same material will float the same way regardless of their size or shape or how much water they use. It is at that point that they are ready to give up their "More A–More B" rule and consider a new explanation.

For Conceptual Change

This is what some constructivists describe as inquiry-based teaching for conceptual change as opposed to just inquiry-based teaching. Students can engage in inquiry, make and share their observations, yet revert to their tenaciously held misconceptions if their prior ideas don't surface and aren't challenged. The first step in this process is to use P-E-O probes, such as "The Floating Logs," that will challenge their existing ideas and lead them to discover for themselves new rules to explain observable phenomena. In the context of inquiry, this is what makes a formative assessment probe an assessment *for* learning as opposed to assessment *of* learning. By eliciting students' prior ideas and providing opportunities for them to test their ideas, experience the cognitive dissonance that happens when their predictions do not match observations, and revise their explanations to fit their observations, formative assessment supports conceptual change teaching and learning in the elementary inquiry-based classroom.

REFERENCES

Driver, R., A. Squires, P. Rushworth, and V. Wood-Robinson. 1994. *Making sense of secondary science: Research into children's ideas.* London: RoutledgeFalmer.

Keeley, P., F. Eberle, and J. Tugel. 2007. *Uncovering student ideas in science, volume 2: 25 more formative assessment probes.* Arlington, VA: NSTA Press.

Stavy, R., and D. Tirosh. 2000. *How students (mis-)understand science and mathematics-intuitive rules.* New York: Teachers College Press.

INTERNET RESOURCE

Uncovering Student Ideas in Science series
   www.nsta.org/publications/press/uncovering.aspx

NSTA CONNECTION

Read the introduction to *Uncovering Student Ideas in Science, Volume 1,* and download a full-size "Floating Logs" probe at *www.nsta.org/SC1010.*

"More A-More B" Rule
Reflection and Study Guide

QUESTIONS TO THINK ABOUT AFTER YOU READ THIS CHAPTER

1. Sink-and-float investigations are commonly found in elementary instructional materials. Why do you think sinking and floating is a good phenomenon for young children to investigate?

2. What does it mean to "develop a rule" to explain a phenomenon in science? Many of the formative assessment probes in the *Uncovering Student Ideas in Science* series ask students to describe the rule or the reasoning they used for their answer. Why might it be better to ask young children to describe a rule rather than ask them to describe their reasoning?

3. Preconceptions are the ideas children bring to their learning based on their everyday experiences and interactions with the natural world. Preconceptions about floating and sinking may develop from experiences with bathtub toys. Can you think of other everyday experiences children have had through play, interactions with the natural world, or prior knowledge that affect how they think about floating and sinking?

4. What is the advantage to using a formative assessment probe such as "Floating Logs" prior to planning instruction that involves floating and sinking?

5. Students use the "More A–More B" intuitive rule to explain many phenomena in science. Can you think of other examples where children use this rule in science? If you have any of the books in the *Uncovering Student Ideas in Science* series, look for other examples of probes that elicit the "More A–More B" intuitive rule.

6. How can you use the P-E-O strategy with the "Floating Logs" probe to support conceptual change through inquiry? How is this different from merely launching into an inquiry investigation?

7. Some educators refer to formative assessment as assessment *for* learning, rather than assessment *of* learning. How does changing the preposition make a difference? What about using a third preposition—assessment *as* learning? How could the "Floating Logs" formative assessment probe be an example of assessment as learning?

8. What are some other examples of More A–More B related to floating and sinking?

9. What does it mean to experience cognitive dissonance? How does cognitive dissonance support learning in science?

10. *A Framework for K–12 Science Education* and *The Next Generation Science Standards* emphasize the scientific and engineering practices as one of the three dimensions for learning science. Which of these practices can be supported through use of the "Floating Logs" probe and how would you use the probe to support each of those practices? (For more information about scientific and engineering practices, visit NSTA's *NGSS* portal at *http://ngss.nsta.org*)

PUTTING FORMATIVE ASSESSMENT INTO PRACTICE

1. How did this probe fit into your curriculum? What was the learning target this probe addressed? What new learning target(s) did you develop for instruction based on the results of this probe?

2. What did you learn about your students' ideas by examining their responses to the probe? Were you surprised by any of their responses?

3. Do any of the "rules" or the reasoning your students used in their explanations reveal the use of the "More A–More B" intuitive rule?

4. Examine the explanations to the correct responses (answer choice B). Do students' explanations provide evidence of complete understanding? Share an example of an explanation that provides no or partial evidence of understanding, even though the student selected the best answer. What can you do to address students' partial understanding?

5. What will you do to address your students' misconceptions about floating and sinking? Describe your next steps in instruction and the instructional opportunities your students will need in order to experience conceptual change.

6. What can you do as a post-assessment to find out how your students' ideas about floating and sinking have changed? How can you find out if they no longer intuitively rely on the "More A–More B" rule?

7. Based on what you learned from using the probe with your students, what suggestions do you have for your colleagues and future teachers?

GOING FURTHER

1. Read and discuss the Teacher Notes for the "Floating Logs" probe (Keeley, Eberle, and Tugel 2007, pp. 28–32). Pay particular attention to the Related Research and Suggestions for Instruction and Assessment sections.

2. To learn more about intuitive rules, consider reading and discussing the book, *How Students (Mis)Understand Science and Mathematics: Intuitive Rules* (Stavy and Tirosch 2000).

3. Read and discuss the teacher notes for the K–2 probe, "Watermelon and Grape" in *Uncovering Student Ideas in Primary Science* (Keeley 2013). How is this probe similar to "Floating Logs"? How is the Related Research for this probe related to the "Floating Logs" probe?

4. Read and discuss the article Smithenry, D. and J. Kim. 2010. Beyond predictions. *Science and Children* 48(2): 48–52.

Chapter 3
Does It Have a Life Cycle?

Life cycles, a common topic in elementary science, help students develop an understanding of the continuity of life. The K–4 National Science Education Standards state that students should know that plants and animals have life cycles that include being born, developing into adults, reproducing, and eventually dying. The details of this life cycle are different for different organisms (NRC 1996). Beginning in the early grades, children have firsthand experiences observing and describing the life cycle of living organisms such as the butterfly, frog, mealworm, or bean plant. Direct experiences observing how an organism changes from egg to adult contribute to an understanding of the cyclic nature of birth, growth, development, and reproduction—concepts that lead to the bigger idea of life continuing from generation to generation. If life continues from generation to generation, then all plants and animals must go through a life cycle, even though it may be different from organism to organism. Is this what students have "learned," or do they have their own private conceptions about life cycles? The formative assessment probe "Does It Have a Life Cycle?" reveals some surprising ideas children have about life cycles (Keeley, Eberle, and Dorsey 2008). Take the time to uncover your students' preconceptions before teaching a unit on life cycles.

Limitations of Context

The "Does It Have a Life Cycle?" probe (Figure 3.1, p. 19) is designed to find out whether students recognize that all living multicellular organisms go through a life cycle. Everything on the list goes through a life cycle, but these cycles vary in length and the type of development juveniles go through. Hundreds of responses from students in grades 1–4 were analyzed from this probe, and more than 80% of the students failed to check off all of the organisms. One of the common reasons given was that an organism has to "go through a change" to have a life cycle. After further probing it was revealed that many of these students considered change to be the type of physical transformation organisms such as frogs and butterflies go through during complete metamorphosis. These students checked off the frog, butterfly, and other insects, but did not check off plants or organisms such as the cow or humans. Was context a factor that inhibited these students' learning? Perhaps the students experienced a butterfly or frog unit, but failed to develop the generalization that all animals, not just frogs and butterflies, have life cycles. The results of this probe can be used to inform teachers of the limitations of context. If you are teaching about life cycles using the butterfly's life cycle, which students can directly observe, then you must be sure to help students understand that the butterfly is one example of an organism that has a particular type of life cycle. All organisms have life cycles, and some of their life cycles may be very different from the butterfly's life cycle.

Using the Probe

This particular type of probe is a form of assessment probe called a "justified list" (Mundry, Keeley, and Landel 2010). This type of probe asks students to check off all the things that fit a particular statement. In this case, all the things that have a life cycle. Students then have to describe the rule or reasoning they used to decide on the examples they picked. It is this part of the probe that allows you to get into your students' heads and examine their thinking. As you uncover reasons for their failure to generalize, the information is used to inform instruction that will address these learning barriers. Make sure the organisms on the list are familiar to students. Remove or replace ones that your students may not recognize. Consider combining words with pictures for younger students or English language learners.

A written assessment in which students check off the things that have a life cycle, explain their reasoning, and turn it in for analysis is one way to administer the probe. As an assessment that promotes learning, a more powerful way to use this probe with young children is through science talk. During "circle time," present one example at a time to the class, by either writing the name of the organism on a chart or holding up a picture of the organism and asking if it has a life cycle—why or why not. As students agree and disagree with each other, carefully listen to their ideas without passing judgment on whether they are right or wrong. Guide the class to come up with a general rule to decide whether an organism has a life cycle, and carefully note the organisms students have difficulty agreeing on. Then, decide on ways to challenge students with their thinking over the course of the lessons you will use to address students' preconceptions.

FIGURE 3.1.

The Does It Have a Life Cycle? Probe

Does It Have a Life Cycle?

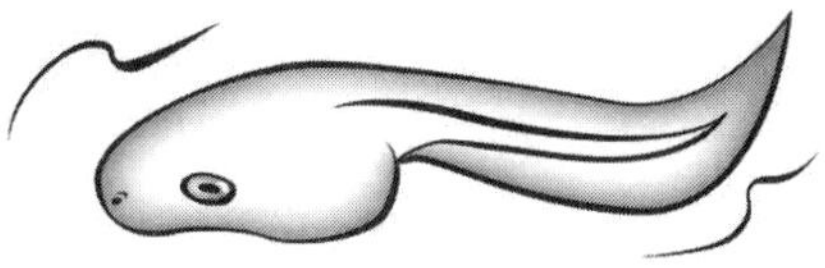

How do you decide if an organism goes through a life cycle? Put an X next to the organisms that have a life cycle.

_____ frog	_____ cow	_____ daisy
_____ butterfly	_____ mushroom	_____ chicken
_____ grasshopper	_____ grass	_____ maple tree
_____ fern	_____ earthworm	_____ human
_____ shark	_____ snail	_____ beetle
_____ bean plant	_____ mold	_____ crab
_____ snake	_____ spider	_____ moth

Explain your thinking. Describe the rule or reason you used to decide if an organism has a life cycle.

Another way to implement the probe is with the formative assessment classroom technique (FACT) called a *card sort* (Keeley 2008). The organisms on the justified list are placed on cards (words or pictures can be used). Students work in small groups to arrange the cards into three groups—things that have a life cycle, things that don't have a life cycle, and things they aren't sure about yet. Students discuss each organism before placing it into a category, justifying their reasons for doing so. If the group cannot agree, the card is placed in the third category and revisited later or during the whole-class discussion. As students are actively discussing their ideas, circulate among the groups, listening carefully to their reasoning and noting the examples that are most problematic. Make stops to probe each group's thinking. For example, say some students were using the cyclic representation to decide which organisms had a life cycle. If they had seen drawings that depicted the stages the organism went through as it developed (e.g., egg, larva, pupa, adult) represented in a circular diagram, they interpreted this as a life cycle. To them, the other organisms lived and died (such as humans) but they didn't have "cycles" because their pictures as they go through life are not shown as a circle diagram.

This influence of a type of representation may not have surfaced without probing the group further. Clearly this is formative information that can be used to make sure representations students see in their textbooks and instructional materials do not contribute to this misconception.

Make It Formative

Remember, a probe is not formative unless you use the data to inform your teaching. One of the ways a justified list probe can inform teaching, not only about life cycles but with many topics in science, is to point out the limitations of context. For students to really learn a concept or idea, they must be able to transfer it to a variety of examples and develop generalizations that apply in a variety of contexts. The "Does It Have a Life Cycle?" probe is one example of a probe that alerts teachers to the need to make sure multiple examples are used during instruction. The butterfly or frog unit is a relevant and engaging context in which to learn about life cycles. However, if the context unintentionally implies that organisms only have life cycles if they go through stages similar to the organism they studied in their life cycle unit, then instruction really has not fully addressed the important learning goal from the National Science Education Standards. Will your students' ideas be limited by the context in which they learned about life cycles? Try out this probe—your students' answers might surprise you and lead you to using other probes to examine the effect of instructional contexts on your students' ideas.

REFERENCES

Keeley, P. 2008. *Science formative assessment: 75 strategies linking assessment, instruction, and learning.* Thousand Oaks, CA: Corwin Press.

Keeley, P., F. Eberle, and C. Dorsey. 2008. *Uncovering student ideas in science, volume 3: Another 25 formative assessment probes.* Arlington, VA: NSTA Press.

Mundry, S., P. Keeley, and C. Landel. 2010. *A leader's guide to science curriculum topic study.* Thousand Oaks, CA: Corwin.

National Research Council (NRC). 1996. *National science education standards.* Washington, DC: National Academies Press.

INTERNET RESOURCE

Uncovering Student Ideas in Science series
www.nsta.org/publications/press/uncovering.aspx

NSTA CONNECTION

Read the introduction to *Uncovering Student Ideas in Science, Volume 1,* and download a full-size "Does It Have a Life Cycle?" probe at *www.nsta.org/SC1011.*

● ●

Does It Have a Life Cycle?
Reflection and Study Guide

QUESTIONS TO THINK ABOUT AFTER YOU READ THIS CHAPTER

1. Life cycles of organisms is a common curricular topic in elementary science. Why is it important for students to learn about life cycles of different organisms?

2. What kinds of instructional experiences have your students had or will have related to life cycles of organisms (these can be in the grade you teach or other grades)? What kinds of real-life experiences or prior knowledge obtained through books, television, or other media do students bring to the classroom about life cycles?

3. What is meant by the limitation of context? Can you think of other examples in which students may fail to develop a broader generalization as a result of being limited by the context in which they learned about the idea or concept?

4. Learning about life cycles in the elementary grades often includes the concept of complete and incomplete metamorphosis. How can learning about this concept muddle students' thinking about life cycles? What can you do to make sure students develop the generalization that all organisms go through a life cycle and that it does not always involve the transformation that organisms like butterflies and frogs experience as part of their life cycle?

5. Justified lists are a type of probe format that is effective in eliciting evidence of the extent to which students can transfer their learning to other examples. If you have any of the books in the *Uncovering Student Ideas in Science* series, look for other examples of justified list probes. Choose one and explain how it can be used to determine if students were limited by their instructional context.

6. Justified list probes such as Does It Have a Life Cycle? can be used in a variety of ways. What are some other ways you can use this probe besides a pencil-and-paper formative assessment?

7. How can the life cycle drawings in instructional materials affect students' understanding of the concept of a life cycle? Search the internet or instructional materials for examples of representations that may impact students' ideas about life

cycles. What are some examples of good representations and examples of representations that may affect students' conceptual understanding of life cycles?

8. When is the use of a probe considered "formative"? What is the difference between diagnostic use of the "Does It Have a Life Cycle?" probe and formative use of the probe? When does a diagnostic probe become formative?

9. *A Framework for K–12 Science Education* and the *Next Generation Science Standards* emphasize crosscutting concepts as one of the three dimensions of science learning. How can the crosscutting concept of Patterns support students' understanding of life cycles? How might you formatively assess how students' connect the concept of patterns to life cycles? (For more information about crosscutting concepts visit NSTA's *NGSS* portal at *http://ngss.nsta.org*)

10. *A Framework for K–12 Science Education* and the *Next Generation Science Standards* emphasize core ideas as dimensions for learning science. These core ideas, combined with practices and crosscutting concepts make up the performance expectation in the *NGSS*. Consider this core idea: Plants and animals have unique and diverse life cycles. This core idea is included in the third-grade *NGSS* performance expectation: 3-LS-1: Develop models to describe that organisms have unique and diverse life cycles but all have in common birth, growth, reproduction, and death. How can you use the "Does It Have a Life Cycle?" formative assessment probe to determine learning targets for the core idea that will prepare students for the performance expectation? What other ways could you formatively assess students' understanding of the life cycle core idea? (For more information about the *NGSS* core ideas and performance expectations, visit NSTA's *NGSS* portal at *http://ngss.nsta.org*)

PUTTING FORMATIVE ASSESSMENT INTO PRACTICE

1. When and where are life cycles addressed in your district's science curriculum?

2. What did you learn about your students' ideas by examining their responses to the probe? Were you surprised by any of their responses?

3. Do any of the students' explanations reveal the limitation of context? Share an example.

4. Did you use a strategy other than pencil and paper with this probe (e.g., card sort, claim cards, group Frayer model)? How did the strategy work? What did you learn about students' thinking by using the strategy with the probe? How did the strategy promote learning? Are there other strategies you might use with this probe (See Keeley 2008, 2014)?

5. What will you do to address your students' misconceptions about life cycles of organisms? Describe your next steps in instruction and the experiences your students will need to develop the generalization that all organisms go through a cycle of birth, growth and development, reproduction, and death.

6. What can you do as a post-assessment to find out how your students' ideas about life cycles have changed?

7. Do you think your students understand the concept of a cycle? What could you do to formatively assess their use of this terminology?

8. Based on what you learned from using this probe with your students, what suggestions do you have for your colleagues and future teachers?

GOING FURTHER

1. Read and discuss the Teacher Notes for the "Does It Have a Life Cycle?" probe (Keeley, Eberle, and Dorsey 2008, pp. 112–116). Pay particular attention to the Related Research and Suggestions for Instruction and Assessment sections.

2. Read and discuss the section on crosscutting concepts on pages 83–85 in *A Framework for K–12 Science Education* (NRC 2012) or online at *www.nap.edu/openbook.php?record_id=13165&page=83.*

3. The concept of a life cycle is included in the *NGSS* and *A Framework for K–12 Science Education* under LS1.B: Growth and Development of Organisms. Read the section on pages 145–147 or online at *www.nap.edu/openbook.php?record_id=13165&page=145* and discuss how the concept of a life cycle is important to developing core ideas about reproduction, growth, and development.

Chapter 4
To Hypothesize or Not?

Formative assessment probes are used not only to uncover the ideas students bring to their learning, they can also be used to reveal teachers' common misconceptions. Consider a process widely used in inquiry science—developing hypotheses. Perhaps you require your students to develop a hypothesis before engaging in the process of inquiry. If so, how well do you (and your students) really understand the meaning and use of a hypothesis? Before reading further, think how you would respond to the probe "What Is a Hypothesis?" (Keeley, Eberle, and Dorsey 2008; Figure 4.1, p. 27). Which of the statements A–N describe a hypothesis? Jot down your answers and think about your conception of what a hypothesis is (and isn't).

What Are Hypotheses?

Dr. Jerry Pine, a scientist from Cal Tech, has worked extensively with K–12 teachers. He wrote in a National Science Foundation monograph: "Frequently the scientific method as taught by nonscientists requires that a scientific inquiry must stem from a hypothesis, which in fact is usually not true" (Institute for Inquiry 1999, p. 61). Many of the great discoveries that stemmed from scientific inquiry were investigations into the unknown. There was insufficient prior knowledge or observations that could support an initial hypothesis that would guide these historic investigations. For example, Charles Darwin did not have a hypothesis in mind when he set out on the *Beagle* and formulated his theory of natural selection.

Which of the probe statements did you choose that matched your idea of what a hypothesis is? Let's compare your answers with the probe's "best choices"—A, B, G, K, L, and M. Hypotheses are tentative explanations that can be tested. They are not, as the popular moniker suggests, an "educated guess." There is no guessing involved. Hypotheses are based on observation or prior knowledge. Hypotheses are not investigative questions, nor are they questions posed at the beginning of an investigation. They are used to investigate a question, guide the investigation, and determine what data to pay attention to. Hypotheses are not predictions. Hypotheses provide a tentative explanation

WHAT ARE THEY THINKING?

that can lead to a prediction. Predictions describe the outcome of an investigation but they do not include an explanation like a hypothesis does.

Not every investigation requires a hypothesis. A lot of science is observational and descriptive and may not involve hypothesis testing. When hypotheses are tested, the results may support or refute a hypothesis, but they never "prove" a hypothesis. Hypotheses are not the first step to becoming a theory. Hypotheses and theories are like apples and oranges—they have different features and a different purpose. Formulating hypotheses is a creative and imaginative endeavor—very different from the stilted hypotheses students develop from cookbook labs. Like the "scientific method," there is no lock-step, prescribed way to develop a hypothesis. It is helpful to frame a testable hypothesis using "if … then," but not all hypotheses require this format.

Now that the best choices and the distracters to this probe have been clarified, has it changed your conception of a hypothesis? Are there certain "caution words" like *prove* that you will be sure your students refrain from using? Will you drop that popular moniker, "educated guess"? Or perhaps you will ask your students to make a prediction instead of a hypothesis because they lack the background knowledge or observations needed to develop a hypothesis. Or maybe it's better to just start with a question. Dr. Jerry Pine said "If we don't begin with a hypothesis, then what does initiate a scientific inquiry? A question … We can begin every scientific investigation with a question. If we insist on a hypothesis, we will often merely force an unscientific guess" (Institute for Inquiry 1999, p. 61).

For Teachers *and* Students

This probe is an example of how formative assessment probes can also be used for teacher learning. After answering the probe yourself and examining the explanation, you may have experienced some cognitive dissonance that will lead to changing your conception of "hypothesis" and more important, changing how you convey this critical term and skill to your students. This is transformative learning—learning that leads to fundamental changes in deeply held beliefs or misconceptions.

This particular probe is designed for middle, high school, and college students. To change it to a probe that is better suited for elementary grades, try the following friendly talk probe:

Three friends were talking about science investigations. This is what they said:

Otis: *I think all investigations should start with a question.*

FIGURE 4.1.

The What Is a Hypothesis? Probe

What Is a Hypothesis?

Hypotheses are used widely in science. Put an X next to the statements that describe a hypothesis.

_____ **A**   A tentative explanation

_____ **B**   A statement that can be tested

_____ **C**   An educated guess

_____ **D**   An investigative question

_____ **E**   A prediction about the outcome of an investigation

_____ **F**   A question asked at the beginning of an investigation

_____ **G**   A statement that may lead to a prediction

_____ **H**   Included as a part of all scientific investigations

_____ **I**   Used to prove whether something is true

_____ **J**   Eventually becomes a theory, then a law

_____ **K**   May guide an investigation

_____ **L**   Used to decide what data to pay attention to and seek

_____ **M**   Developed from imagination and creativity

_____ **N**   Must be in the form of "if…then…"

Describe what a hypothesis is in science. Include your own definition of the word *hypothesis* and explain how you learned what it is.

_____

_____

_____

_____

Jamie: *I think all investigations should start with a hypothesis.*

Anna: *I think all investigations should start with a prediction.*

Which friend do you most agree with and why? Explain your thinking about starting investigations.

The best answer is Otis. As Dr. Pine would say, "To hypothesize or not to hypothesize? Don't. Pose a question instead (Institute for Inquiry 1999, p. 61).

REFERENCES

Institute for Inquiry. 1999. *Foundations: Inquiry—thoughts, views, and strategies for the K–5 classroom.* Washington, DC: National Science Foundation.

Keeley, P., F. Eberle, and C. Dorsey. 2008. *Uncovering student ideas in science, volume 3: Another 25 formative assessment probes.* Arlington, VA: NSTA Press.

INTERNET RESOURCE

Uncovering Student Ideas in Science series
www.nsta.org/publications/press/uncovering.aspx

NSTA CONNECTION

Read the introduction to *Uncovering Student Ideas in Science, Volume 1,* and download a full-size "What Is a Hypothesis?" probe at *www.nsta.org/SC1012.*

• •

To Hypothesize or Not
Reflection and Study Guide

QUESTIONS TO THINK ABOUT AFTER YOU READ THIS CHAPTER

1. Think about how you would have answered this probe prior to reading this chapter. How did your K–12 science education influence your understanding of the meaning and use of a hypothesis in science?

2. What are some of the ideas about hypotheses that surprised you after reading this article?

3. Why do you think a hypothesis is often defined as an educated guess? Why is it a good idea to avoid using this definition when teaching the concept and use of a hypothesis?

4. Think about your own teaching. When your students develop and test hypotheses, how does their use of hypotheses match the descriptions in the article?

5. What is meant by a caution word? Can you think of other caution words to be aware of when teaching science?

6. Think of an example in your teaching when it was appropriate for students to develop a hypothesis. Think of an example where it would be appropriate for them to make a prediction, rather than develop a hypothesis. Now think of an example where developing a question would be more appropriate than making a prediction or developing a hypothesis.

7. Think of an example where it is appropriate to use an "if … then …" format to state a hypothesis. Now think of an example where an "if … then …" format is not used to state a hypothesis.

8. Examine your curriculum and instructional materials. How do your curriculum and materials address the development or use of hypotheses? How do your curriculum and instructional materials reflect the information about hypotheses in this chapter?

9. After reading this chapter, how would you define the word *hypothesis*?

10. *A Framework for K–12 Science Education* and the *Next Generation Science Standards* emphasize scientific and engineering practices as one of the three dimensions of science learning. How does the practice of planning and carrying out investigations include the use of hypotheses? At what grade level are students expected to develop and use hypotheses? (For more information about the scientific and engineering practices visit NSTA's *NGSS* portal at *http://ngss.nsta.org*.)

PUTTING FORMATIVE ASSESSMENT INTO PRACTICE

1. What did you learn about your students' use of the word *hypothesis* after examining their responses to the probe? Were you surprised by any of their responses?

2. Do any of their responses indicate where their ideas about hypotheses came from? How is knowing the origin of their interpretation of hypothesis helpful in making formative decisions about next steps for instruction?

3. What will you do next to address your students' misunderstandings about hypotheses?

4. What modifications will you make to your curriculum, instruction, or instructional materials to ensure students develop an accurate understanding of what a hypothesis is and how it is used?

5. What can you do as a post-assessment to find out how your students' ideas about hypotheses have changed?

6. Based on what you learned from using the probe with your students, what suggestions do you have for your colleagues and future teachers?

GOING FURTHER

1. Read and discuss the Teacher Notes for the "What Is a Hypothesis?" probe (Keeley, Eberle, and Dorsey 2008, pp. 102–108). Pay particular attention to the Related Research and Suggestions for Instruction and Assessment sections.

2. Read and discuss the full article by Jerry Pine (Institute for Inquiry 1999). You can access this NSF monograph online at *www.nsf.gov/pubs/2000/nsf99148/ch_7.htm*. Scroll down to the end of Chapter 7 to find the section "To Hypothesize or Not to Hypothesize."

3. Read and discuss the section on Planning and Carrying Out Investigations on pages 59–61 in *A Framework for K–12 Science Education* (NRC 2012) or online at *www.nap.edu/openbook.php?record_id=13165&page=59*.

4. Read and discuss the article Baxter, L., and M. Kurtz. 2001. When a hypothesis is not an educated guess. *Science and Children* 38 (8): 18–20.

5. Watch and discuss the NSTA archived web seminar on the practice of planning and carrying out scientific investigations *http://learningcenter.nsta.org/products/symposia_seminars/NGSS/webseminar7.aspx*

Chapter 5
How Far Did It Go?

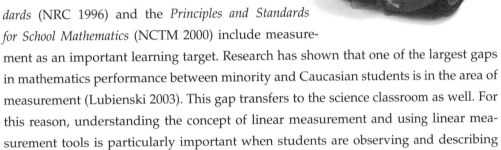

At the elementary level, measurement is a concept and process common to both science and mathematics. Both the *National Science Education Standards* (NRC 1996) and the *Principles and Standards for School Mathematics* (NCTM 2000) include measurement as an important learning target. Research has shown that one of the largest gaps in mathematics performance between minority and Caucasian students is in the area of measurement (Lubienski 2003). This gap transfers to the science classroom as well. For this reason, understanding the concept of linear measurement and using linear measurement tools is particularly important when students are observing and describing the motion of objects.

Inquiry provides an implicit opportunity for students to apply their understandings of measurement units and the tools used to measure the change in the position of an object after it has traveled a distance. However, without explicit, direct instruction in measurement and the instruments of measurement that takes into account students' common misunderstandings, significant errors in measurement may develop early in elementary grades that continue to affect students' ability to use these important concepts and processes in successive grades.

Using the Probe

The formative assessment probe "How Far Did It Go?" (Figure 5.1, p. 33) in *Uncovering Student Ideas in Physical Science, Volume 1: 45 Force and Motion Assessment Probes* (Keeley and Harrington 2010) can be used to reveal whether students recognize that units of distance traveled must be measured from the starting point to the ending point. It is especially useful in determining how students measure length when there is a nonzero origin. Student responses reveal a common error pattern that children make in both science and mathematics. Common error patterns refer to the systematic use of inaccurate and inefficient procedures or strategies (Rose Tobey and Minton 2010). For example, a common error pattern that applies to this motion assessment probe is the consistent misreading of a measurement device.

Research in mathematics has revealed that few children recognize that any point on a measurement scale can serve as a starting point. Studies of children, all the way up through grade 5, have shown they have the tendency to read whatever number is at the end point. In the case of this probe, many students selected E: 10 units as their answer. When asked how they figured out their answer, the most common response was "The front wheels of the car ended on the 10 mark." These students did not take into account the nonzero starting point.

The best answer is C: 6. Notice how the back of the car is positioned at the "2" mark. After the car moves and then stops, the back of the car is positioned at the "8" mark. The car moved 6 units between the "2" mark and the "8" mark. Students get the same result if they measure from the front of the car. The front of the car is initially positioned at the "4" mark. After traveling, the front of the car stops at the "10" mark, resulting in a distance traveled of 6 units. However, some students take into account the starting point of the car but are not consistent in using the same part of the car to identify the ending point. For example, students who choose answer D: 8 begin at the back of the car on the "2" starting point and end with the front of the car at the "10" ending point, resulting in a distance traveled of 8 units. Conversely, students who chose answer B: 4 begin at the front of the car with a starting point of "4" and end at the back of the car with an ending point of "8" for a total of 4 units traveled. And finally, some students select answer A: 2 by measuring the length of the car, rather than the distance the car traveled. These students may not understand what distance means.

Teaching Implications

Responses to this probe clearly indicate the need for explicit teaching in both science and mathematics about starting and ending points when using linear measurement tools. It also points out the importance of providing opportunities for students to measure different types of objects and motions using different starting points. Students need opportunities to measure the same object and motion when the length or distance are the same but the starting points differ, so that students see that the length or distance stays the same regardless of starting point. The probe can be modified to determine whether students respond differently when the car starts at a zero or the "1" mark. For students who merely read off the end point without considering the starting point, a useful strategy might be to help them differentiate between *where* an object ends up from *how far* the object has gone.

FIGURE 5.1.

The How Far Did It Go? Probe

How Far Did It Go?

Before the car moves

After the car moves and stops

Gracie wants to measure the distance that her toy car travels. She places her car next to a measuring tape as shown in the first picture. She pushes the car. The second picture shows how far Gracie's car traveled until it stopped. Gracie measures the distance her car moved.

Circle the number of measurement units that best describes how far Gracie's car moved.

A 2

B 4

C 6

D 8

E 10

Describe how you figured out your answer.

_____

_____

_____

_____

_____

_____

Consider combining this probe, which is used in a motion unit, with similar types of diagnostic and formative assessment that are developed for mathematics such as those in *Uncovering Student Thinking in Mathematics K–5: 25 Formative Assessment Probes for the Elementary Classroom* (Rose Tobey and Minton 2010). Combined, both of these "Uncovering" books show the link between similar conceptual and procedural misunderstandings in both mathematics and science and support the integration of important topics like measurement.

For example, mathematics educators point out that length is usually the first attribute children measure in mathematics; however, length measurement is not well understood by young children. There is a strong temptation to explain to students how to use units and devices to measure length and then send them off to practice measuring. The attention shifts from developing a conceptual understanding of measurement using units to one that is merely procedural (Rose Tobey and Minton 2010). Does this sound familiar in science? Rose Tobey's book includes several excellent K–5 probes that elicit students' understanding of length measurements as a result of matching a length with a number of units rather than as a number on a ruler or measurement device. As you use these measurement probes, it will be obvious to you that we cannot take for granted that children learn measurement merely by providing opportunities to measure. You must take the time to use carefully designed probes and watch, listen to, and determine the procedures and strategies children use to measure length or distance. The data will help you consider the teaching implications and make adjustments to your instruction based your examination of student thinking that results from using these probes.

REFERENCES

Keeley, P., and R. Harrington. 2010. *Uncovering student ideas in physical science, volume 1: 45 force and motion assessment probes.* Arlington, VA: NSTA Press.

Lubienski, S. 2003. Is our teaching measuring up? Race-, SES-, and gender-related gaps in measurement achievement. In *Learning and teaching measurement: 2003 yearbook,* eds. D.H. Clements and G. Bright, 282–292. Reston, VA: National Council of Teachers of Mathematics.

National Council of Teachers of Mathematics (NCTM). 2000. *Principles and standards for school mathematics.* Alexandria, VA: NCTM.

National Research Council (NRC). 1996. *National science education standards.* Washington DC: National Academies Press.

Rose Tobey, C., and L. Minton. 2010. *Uncovering student thinking in mathematics K–5: 25 formative assessment probes for the elementary classroom.* Thousand Oaks, CA: Corwin Press.

INTERNET RESOURCE

Uncovering Student Ideas in Science series
  www.nsta.org/publications/press/uncovering.aspx

NSTA CONNECTION

Read the introduction to *Uncovering Student Ideas in Science, Volume 1,* and download a
full-size "How Far Did It Go?" probe at *www.nsta.org/SC1101.*

• •

How Far Did It Go?
Reflection and Study Guide

QUESTIONS TO THINK ABOUT AFTER YOU READ THIS CHAPTER

1. Measurement is an area that has one of the largest performance gaps in mathematics. Do you think measurement is an area where there is also a large performance gap in science?

2. What are some ways your students use linear measurement and linear measurement tools in science?

3. Examine each of the answer choices to the "How Far Did It Go?" assessment probe. What common error or misunderstanding is represented by each of the answer choices?

4. What does it mean to be explicit when teaching measurement? What are the kinds of opportunities children need in order to understand measurement both conceptually and procedurally?

5. What do you think about the statement, "We cannot take for granted that children learn measurement merely by providing opportunities to measure." How do your students currently learn about measurement?

6. What do you think is the difference between understanding length measure as the result of matching a length with a number of units versus matching a length with a number on a ruler?

7. What are some other examples of science investigations in which students may need to measure from a nonzero starting point?

8. How are the difficulties students have with measurement in the mathematics classroom similar to the difficulties they have with measurement in the science

classroom? What implications does this have for STEM teaching and learning or cross-curricular integration?

9. At first glance this probe may look like it is designed for early elementary grades (K–2). Do you think students in the intermediate grades (3–5) and middle school (6–8) also struggle with linear measurement? When and why would you use this probe?

10. *A Framework for K–12 Science Education* and the *Next Generation Science Standards* emphasize scientific and engineering practices as one of the three dimensions of learning science. Two of these practices are planning and carrying out investigations and the use of mathematics and computational thinking. How does this probe support those scientific and engineering practices (For more information about scientific and engineering practices visit NSTA's *NGSS* portal at *http://ngss. nsta.org*)?

PUTTING FORMATIVE ASSESSMENT INTO PRACTICE

1. What did you learn about your students' understanding of linear measurement after using this probe? Were you surprised by any of their answer choices and explanations?

2. It has been suggested that opportunities to work with "broken rulers," having students construct their own measurement tools, and experiencing a variety of standard and nonstandard measuring devices can help students develop a better understanding of measurement. Based on your students' responses to this probe, would any of these suggestions be appropriate for your students?

3. What modifications will you make to your instruction or instructional materials to ensure students develop a conceptual and procedural understanding of linear measurement that they can apply in a variety of science contexts?

4. Do the results from this probe indicate any gaps in your science or mathematics curriculum or learning targets related to measurement? What will you do to address these gaps?

5. Based on what you learned about your students' common difficulties and misconceptions related to measurement, what suggestions do you have for your colleagues and future teachers?

GOING FURTHER

1. Read and discuss the Teacher Notes for the "How Far Did It Go?" probe (Keeley and Harrington 2010, pp. 16–17). Pay particular attention to the Related Research and Suggestions for Instruction and Assessment sections.

2. Read and discuss the sections on the scientific and engineering practices of planning and carrying out investigations on pages 59–61 and using mathematics and computational thinking on pages 64–67 in *A Framework for K–12 Science Education* (NRC 2012) or online at *www.nap.edu/openbook.php?record_id=13165&page=59*. How is measurement included as part of these practices?

3. *The Common Core State Standards for Mathematics* includes standards for Measurement and Data. Examine the Measurement and Data standards for your grade level at *www.corestandards.org/math* (or examine your own state standards). Examine and discuss how the core ideas related to measurement in mathematics support the use of measurement in science.

4. Read and discuss the article on linear measurement in science: Smith, L., D. Sterling, and P. Moyer-Packenham 2006. Activities that really measure up. *Science and Children*. 44 (2): 30–33.

Chapter 6
Needs of Seeds

Seeds provide a context for many elementary inquiry-based activities that address ideas related to the needs of organisms and life cycles. The "Needs of Seeds" formative assessment probe (Figure 6.1, p. 41) can be used to find out whether students recognize that seeds have needs both similar to and different from plants and other living organisms (Keeley, Eberle, and Tugel 2007). The probe reveals whether students overgeneralize the needs of seeds by assuming they have the same needs as the adult plants that grow and develop from a seed during the life cycle of the plant. Through a process of elicitation, exploring, and testing their ideas about seeds, engaging in argumentation and sense making, and reflecting back on their initial thinking, "Needs of Seeds" provides an assessment springboard for students to learn about the needs of living things, hone their skills of inquiry, and experience the nature of science. In addition, teachers can gather valuable formative assessment information about their students' thinking and their ability to identify the evidence they need to support their ideas.

Probe Preparation

Because probes are designed to elicit students' preconceptions, they are carefully constructed to avoid terminology that students may not yet have had an opportunity to learn. For example, this probe targets the concept of germination and what a seed needs to germinate. Students who encounter the word *germinate* without understanding its meaning may respond differently to the probe if they confuse the word with the growth of a young plant. Therefore, the probe intentionally uses the word *sprout* because it has a more familiar meaning to students. If students' scientific vocabulary includes the word *germination*, then it is appropriate to add this word to the probe and refer to it on a word wall. However, either way, it is important to make sure students know what a "sprouted" or "germinated" seed is before asking them to share their ideas related to the probe.

When possible, try to use props when setting the context for a probe. For example, before students record or discuss their ideas about what a seed needs to sprout, show the students two seeds—a recently sprouted seed and an unsprouted seed. (Germinating dried lima beans inside a wet paper towel works well for this demonstration but make sure students did not see how you germinated the seed as it will affect their response to the probe.) Have students carefully examine the difference in the two seeds, noting that the sprouted seed has a tiny plant just beginning to emerge from the seed. Emphasize that this is what it means to "sprout." Then students can be asked what seeds need to sprout.

Using the Probe

The "Needs of Seeds" probe is written in a format called a justified list (Mundry, Keeley, and Landel 2010). A justified list probe asks students to select examples from a list that fit a particular criterion and then justify the "rule" or reasoning they used. In this example, students have to select the things that seeds need to sprout. Depending on the grade level of the students, the teacher may use all of the examples listed for the probe, add additional examples, or exclude examples that may be unfamiliar to students.

Although the process of getting seeds to germinate for some species can be complex, the best response is that seeds need water, air, food, and warmth. The details of the biological processes that happen inside the cells of the seed are beyond the elementary level, but a basic explanation is as follows: Water is needed by all living things. Water is absorbed by the seed and used by the cells of the embryo to activate its metabolism and start growing. Oxygen in the air is used for cellular respiration that releases the energy from the food that the seed uses to develop into a seedling. The food contained inside the seed is called the *endosperm*. It is the fleshy part of the seed that provides a source of energy and building material for the growing embryo and the seedling until the plant's first set of true leaves begin to photosynthesize and make food for the plant. Chemical reactions inside the embryo's cells require a warm temperature to support the growth and development of the seed's embryo. However, there are some plants like the red oak that require their seeds to go through a period of cold before they can germinate under warm conditions.

As you use this probe, you may find students confuse the needs of a plant with the needs of the seed from which it came. For example, some students will choose soil and sunlight as a result of their experiences growing plants. Some will draw upon their experiences sowing seeds and choose soil and darkness as a result of planting seeds in a garden. However, seeds can germinate without soil as long as the conditions provide enough moisture and they can sprout in the sunlight as well as in darkness.

After the Probe

One formative assessment classroom technique (FACT) that can be combined with this probe is the A&D Statements technique (Keeley 2007). This technique provides an oppor-

FIGURE 6.1.
The Needs of Seeds Probe

Needs of Seeds

Seeds sprout and eventually grow into young plants called seedlings. Put an X next to the things you think a seed needs in order for it to sprout.

___ water

___ soil

___ air

___ food

___ sunlight

___ darkness

___ warmth

___ Earth's gravity

___ fertilizer

Explain your thinking. Describe the "rule" or reasoning you used to decide what a seed needs in order to sprout.

tunity for students to further elaborate their thinking and decide what type of evidence they could use to further support their ideas. The FACT asks students whether they agree or disagree with the filled-in statement (e.g., Seeds need <u>water</u> to sprout). They can choose the option "it depends on" if it sometimes depends on other factors. For example, some seeds need a period of cold before they can sprout, or seeds need water but too much water will drown them. Students then describe how they can support their ideas through designing experiments or other types of first- or secondhand inquiry.

Be sure to provide an opportunity for the children to engage in discussion in small groups and as a class about each of the statements and have them share why they agree, disagree, or chose "it depends on." A&D statements also reveal to the teacher how students think about pursuing inquiry into questions they wonder about and ways to support their ideas. They may describe how to conduct a "fair test," search for information on the internet, select trade books they could read to gather information, identify an expert they could interview, and so on. It is this type of formative assessment that not

FIGURE 6.2.

A&D Statement for Needs of Seeds

Seeds need _____ to sprout	What do you think?	How can you find out?
water	___ I agree ___ I disagree ___ It depends on:	
soil	___ I agree ___ I disagree ___ It depends on:	
air	___ I agree ___ I disagree ___ It depends on:	
food	___ I agree ___ I disagree ___ It depends on:	
sunlight	___ I agree ___ I disagree ___ It depends on:	
darkness	___ I agree ___ I disagree ___ It depends on:	
warmth	___ I agree ___ I disagree ___ It depends on:	
Earth's gravity	___ I agree ___ I disagree ___ It depends on:	
fertilizer	___ I agree ___ I disagree ___ It depends on:	

only reveals students' preconceptions but also becomes an assessment for learning as students launch into inquiry to "find out." It helps students think about multiple ways to seek answers to questions that sometimes involve testing their ideas and other times involve alternative ways of gathering information. Figure 6.2 is an example of an A&D Statements technique that can be used with this probe. It can be modified for younger children to include the conditions they can easily test and are familiar with such as water, soil, sunlight, darkness, and warmth.

After students have had opportunities to test and research their ideas about the needs of seeds, provide students with an opportunity to revisit their initial ideas and describe how many of their ideas may have changed in light of new evidence. Have them compare the needs of seeds with the needs of a growing plant to see where there are similarities and differences, and how these similarities and differences may have affected their initial thinking. This reflective part of formative assessment is critical in supporting conceptual change. When students recognize the new evidence that led to a change in their initial ideas, this not only helps solidify the concepts learned but also provides a rich opportunity for them to experience the nature of science. By seeing how ideas change when new evidence is presented, discussed, and eventually accepted by the scientific community (in this case their classmates), children are experiencing the way science is carried out in the real world.

Formative assessment holds an amazing potential to grow new ways of informing teaching and learning. Try this assessment probe and the accompanying FACT and see how new ideas and ways of thinking "sprout" in your classroom.

REFERENCES

Keeley, P. 2007. *Science formative assessment: 75 practical strategies for linking assessment, instruction, and learning.* Thousand Oaks, CA: Corwin Press.

Keeley, P., F. Eberle, and J. Tugel. 2007. *Uncovering student ideas in science, volume 2: 25 more formative assessment probes.* Arlington, VA: NSTA Press.

Mundry, S., P. Keeley, and C. Landel. 2010. *A leader's guide to science curriculum topic study.* Thousand Oaks, CA: Corwin Press.

NSTA CONNECTION

Read the introduction to *Uncovering Student Ideas in Science, Volume 1,* and download a full-size "Needs of Seeds" probe at *www.nsta.org/SC1102.*

Needs of Seeds
Reflection and Study Guide

QUESTIONS TO THINK ABOUT AFTER YOU READ THIS CHAPTER

1. Seeds provide a context for a variety of science learning opportunities in elementary science. What are some ways you can use seeds as a context for learning important ideas and practices of science?

2. What is meant by overgeneralization? How can the "Needs of Seeds" probe reveal students' tendency to overgeneralize in science?

3. What does it mean to elicit students' preconceptions? How is the "Needs of Seeds" probe used for elicitation?

4. Why is it important to avoid technical terminology when using a probe for elicitation? Can students use vocabulary in science without conceptual understanding? Can you elicit evidence of conceptual understanding without using technical terminology? When is it appropriate to introduce and use science vocabulary?

5. What is the advantage to using props when introducing a probe? Can you think of other probes you have used that could be introduced using props?

6. The justified list format is frequently used with probes in the *Uncovering Student Ideas in Science* series. How would you describe this format? What do you like about this format?

7. Are there examples you might remove from the "Needs of Seeds" probe? What other things could you add to the list of answer choices for this probe?

8. A&D Statements is an example of a FACT (formative assessment classroom technique). What additional formative information about the needs of seeds does this FACT provide? How could you use this FACT with other science topics?

9. Why is it important for students to have an opportunity to revisit their initial ideas after instruction? How would you use assessment probes for both pre- and postassessment, including reflection?

10. *A Framework for K–12 Science Education* and *The Next Generation Science Standards* emphasize scientific and engineering practices as one of the three dimensions of learning science. One of these practices is engaging in argument from evidence. How does this probe support the use of this scientific practice? (For more information about scientific and engineering practices visit NSTA's *NGSS* portal at *http://ngss.nsta.org*.)

PUTTING FORMATIVE ASSESSMENT INTO PRACTICE

1. What did you learn about your students' understanding of seeds and what seeds require for germination? Did any of your students' answer choices surprise you?

2. What evidence did you find in your students' responses that indicate confusion between the needs of seeds and the needs of plants?

3. What are your next steps for instruction that will address students' ideas about what seeds need to sprout?

4. What modifications will you make to your curriculum, instruction, or instructional materials based on students' responses to this probe?

5. Based on what you learned about your students' ideas about the needs of seeds, what advice or suggestions do you have for your colleagues and future teachers?

GOING FURTHER

1. Read and discuss the Teacher Notes for the "Needs of Seeds" probe (Keeley, Eberle, and Tugel 2007 pp. 102–106). Pay particular attention to the Related Research and Suggestions for Instruction and Assessment sections.

2. Read the Teaching Through Tradebooks article: Ansberry, K. and E. Morgan. 2009. Secrets of seeds. *Science and Children* 46 (5): 161–168.

3. Watch and discuss the NSTA archived web seminar on the *NGSS* scientific practice of engaging in argument from evidence at *http://learningcenter.nsta.org/products/symposia_seminars/NGSS/webseminar11.aspx*.

Chapter 7
The Mitten Problem

I f Deb O'Brien had begun her lesson on heat in the usual way, she might never have known how nine long Massachusetts winters had skewed her students' thinking. Her fourth graders would have learned the major sources of heat, a little bit about friction, and how to read a thermometer. By the end of two weeks, they would have been able to pass a simple test on heat. But their preconceptions, never having been put on the table, would have continued, coexisting in a morass of conflicting ideas about heat and its behavior" (Watson and Konicek 1990, p. 680). In 1992 I read an article about this lesson in *Phi Delta Kappan* that transformed my teaching and to which I credit today, the genesis of my popular formative assessment series, *Uncovering Student Ideas in Science*. The article was titled *Teaching for Conceptual Change: Confronting Children's Experience*. The students were convinced that if they put thermometers in their hats, coats, sweaters, or mittens, the temperature would rise. Not only did they believe the temperature would rise, but they were sure it would go "way up" and get "wicked hot." The students were surprised to find the temperature had not changed when they removed the thermometers. They decided the thermometers were not left in their "warm clothes" long enough. After all, one student explained, when the doctor takes your temperature, you have to leave the thermometer in your mouth a long time. Each time the students found the thermometer reading did not go up, and each time they substituted a new theory to explain their idea that "warm clothes" warm us up.

The teacher refrained from telling them the difference between objects that keep heat in (insulators) from objects that emit heat. She kept probing, giving them opportunities to test their naïve ideas, and had them confront the inconsistencies in their theories. The testing and discussion went on for three days. Finally the children reached an impasse and had no alternative theory to replace the ones they had. The teacher now stepped in and offered them two choices: (1) their theory that heat can come from any object, including hats and mittens, and (2) a new theory she offered that heat can come from warm things, like our bodies, and become trapped inside objects like hats and mittens. She asked the students to write about what they now believed and their reasons for their beliefs. She then asked them to stand in one of the back corners of the room that best matched their thinking: theory 1 or theory 2, or stand in the front if they were unsure.

One by one the children moved to different areas of the room, sometimes changing their mind and moving to a different area as they thought through their ideas.

Instead of telling the students that theory 2 is the correct explanation, the teacher asked the students who chose this new explanation to think about how they could test this new theory. The children came up with the idea of putting thermometers inside their hats, which they wore on their heads, when they went out to recess. And so they did and the inquiry continued. Well, this is a short summary of a remarkable lesson, and I encourage you to read the whole article. What struck me as a teacher at that time was that this was more than just what I knew as inquiry-based teaching, this was inquiry teaching for conceptual change—a constructivist form of inquiry! It still posed investigable questions, involved students in collecting and analyzing data, encouraged explanations and communication, and eventually led students to a deeper understanding of the phenomenon. The teacher artfully designed the inquiry experience by first taking into account her students' preconceptions. Often this important first step is missed as teachers launch into inquiry. She then made sure the inquiry provided an opportunity to confront them with their ideas as they tested their predictions until eventually some students were willing to embrace a new explanation when their old one no longer worked for them.

After reading that article and delving deeper into learning about conceptual change, I decided to start each inquiry lesson with a probing question that would reveal alternative ideas my students might bring to their learning experience. I decided I could use these alternative ideas as springboards into inquiry. I wrote the "Mitten Problem" (Figure 7.1) back in 1990 after reading this article and sure enough, I got similar responses from my eighth graders as the teacher in the article did with her fourth graders. I let my students test their prediction and muddle through with uncertainties until they were ready to accommodate a new explanation. I felt energized by seeing how I could create a community of learners who felt comfortable defending their ideas and arguing with others as they held on tenaciously to their explanations or embraced new ones.

Challenge Conceptions

The "theory of immaculate insulation" remains prevalent among students. Research indicates that students often believe that some materials and objects, such as blankets or mittens, are intrinsically warm. Likewise, they believe that some objects and materials, such as metals, are cold (Driver et al. 1994).

One effective way to gain insight into how students understand concepts in order to facilitate conceptual change in the classroom is to investigate just the opposite—how do

---

FIGURE 7.1.

The Mitten Problem Probe

The Mitten Problem

Sarah's science class is investigating heat energy. They wonder what would happen to the temperature reading on a thermometer if they put the thermometer inside a mitten.

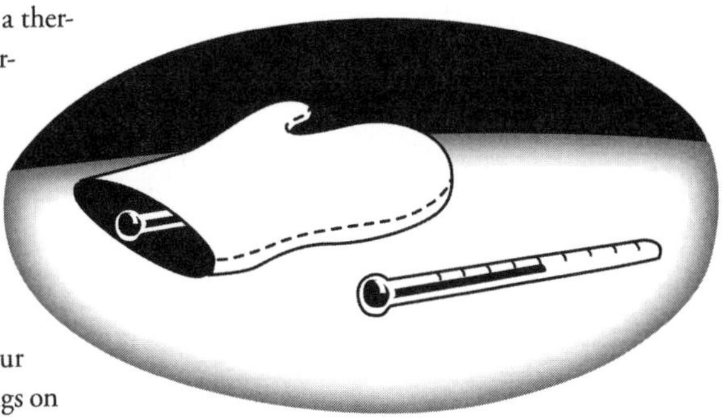

Sarah's group obtained two thermometers and a mitten. They put one thermometer inside the mitten and the other thermometer on the table next to the mitten. An hour later they compared the readings on the two thermometers. The temperature inside the room remained the same during their experiment.

What do you think Sarah's group will discover from their investigation? Circle the response that best matches your thinking.

A The thermometer inside the mitten will have a lower temperature reading than the thermometer on the table.

B The thermometer inside the mitten will have a higher temperature reading than the thermometer on the table.

C Both thermometers will have the same temperature reading.

Describe your thinking. Provide an explanation for your answer.

---

their misunderstandings develop? Where does this idea that objects such as mittens generate their own heat come from? It's quite reasonable to think this when we think about our everyday experiences. What do we do when we are cold? We put on a sweater, mittens, or cover ourselves with a blanket. We use language like "warm clothes" or "warm up your hands with these mittens." Using probes such as the "Mitten Problem" can reveal these tenaciously held ideas that come from children's everyday experiences and make sense to them. (The teacher notes that accompany a probe also provide summaries from the cognitive research that explain the commonly held ideas students have related to a concept.)

Probes such as the "Mitten Problem" are what I call a P-E-O probe: Predict, Explain, Observe. Students commit to an outcome, provide an explanation for their prediction of that outcome, and test their ideas by making observations. When their observations do not match their prediction, new explanations (and sometimes more testing) are needed. You can find many P-E-O probes in the *Uncovering Student Ideas in Science* series to use to spark inquiry in your classroom and facilitate conceptual change in the process.

Since the publication of "The Mitten Problem," I have had the pleasure to get to know and work with one of the authors of the transformative article, *Teaching for Conceptual Change,* Dr. Dick Konicek-Moran. Dick is a retired professor emeritus from the University of Massachusetts and is a natural teacher. I often refer to him as my "muse" since his seminal article (which is now used in many teacher preparation programs as well as the Exploratorium's Institute for Inquiry) led me down the path to develop formative assessment probes. Furthermore, Dick has published an outstanding resource through NSTA Press for elementary teachers that also promotes conceptual change through inquiry and uncovers many of the commonly held ideas noted in the research literature (see Internet Resource). The *Everyday Science Mysteries* series uses a short story genre to provide an opportunity for students to explore their ideas and work toward conceptual change as they take ownership in the inquiry. The formative assessment probes and mystery stories complement each other and can be used together. For example, the story *Warm Clothes?* (Volume 3) provides an opportunity for children to finish the story by testing their ideas about whether warm clothes generate heat. Together these resources will help you facilitate inquiry that encourages students to look for inconsistencies between what they believe they will find out and what their observations actually reveal to them. Some students (and teachers) will find this frustrating and will continue to defend their strongly held ideas. For others, the experience will change their thinking, including the way teachers think about teaching and learning! Try the probe, the mystery story, and read the article. I hope you, too, will be transformed.

REFERENCES

Driver, R., A. Squires, P. Rushworth, and V. Wood-Robinson. 1994. *Making sense of secondary science: Research into children's ideas.* London: RoutledgeFalmer.

Keeley, P., F. Eberle, and L. Farrin. 2005. *Uncovering student ideas in science, volume 1: 25 formative assessment probes.* Arlington, VA: NSTA Press.

Watson, B., and R. Konicek. 1990. Teaching for conceptual change: Confronting children's experience. *Phi Delta Kappan* 71 (9): 680–684. Available online at *www.exploratorium.edu/ifi/resources/workshops/teachingforconcept.html*

INTERNET RESOURCE

Everyday Science Mysteries
www.nsta.org/publications/press/mysteries.aspx

NSTA CONNECTION

Read the introduction to *Uncovering Student Ideas in Science, Volume 1,* and download a full-size "The Mitten Problem" probe at *www.nsta.org/SC0311.*

• •

The Mitten Problem
Reflection and Study Guide

QUESTIONS TO THINK ABOUT AFTER YOU READ THIS CHAPTER

1. How would you have answered this probe before reading the chapter? What preconceptions may have lingered with you since childhood that influenced your thinking about this probe?

2. What do you think about Deb O'Brien's (the teacher in the article) instruction? How is it similar to or different from your teaching?

3. How would the lesson, and as a result the learning, have been different if the teacher corrected the students' misconception that the clothes "warmed" the thermometer? Do you think it is OK to not immediately correct a misconception? When and for how long would you let a misconception linger?

4. How is "inquiry for conceptual change" different from "inquiry"? Do you think "hands-on" is always the best way to learn science?

5. What kind of classroom climate is needed in order to teach for conceptual change? What can you do to create an environment in which students feel safe to share their thinking, regardless of whether they are right or wrong?

6. The author explains how her teaching was transformed after reading the seminal article Teaching for Conceptual Change: Confronting Children's Experience. Have you ever read an article or book that transformed your teaching or beliefs about learning?

7. The author states that misconceptions are not a bad thing when they are used as springboards for learning. Do you think of misconceptions in a negative or positive way?

8. "The Mitten Problem" is an example of a probe that can be used with the P-E-O FACT (formative assessment classroom technique) (Keeley 2008, 2014). How would you use the P-E-O technique with this probe? What other ways could you use the P-E-O technique?

9. This probe is an example of how everyday experiences and the ways we use language can affect the way we think about phenomena. What kinds of everyday experiences and words we use affect students' ideas about heat and temperature?

10. *A Framework for K–12 Science Education* and *The Next Generation Science Standards* emphasize scientific and engineering practices as one of the three dimensions of learning science. One of these practices is engaging in argument from evidence. How does this probe support the use of this scientific practice? (For more information about scientific and engineering practices visit NSTA's *NGSS* portal at *http://ngss.nsta.org.*)

PUTTING FORMATIVE ASSESSMENT INTO PRACTICE

1. What did you learn about your students' understanding of heat and temperature? Were you surprised by any of their answer choices and explanations?

2. How would you categorize your students' reasoning? What are the different ideas they are using to support their answer choices? Where do you think their ideas came from? Do their responses reveal evidence of the source of their misconception or their understanding?

3. What modifications will you make to your curriculum, instruction or instructional materials based on students' responses to this probe?

4. Did you follow up with the P-E-O strategy? If so, what happened after your students had an opportunity to test their ideas? Was their experience similar to the experience of the children in the article?

5. Based on what you learned about your students' ideas about the thermometer in the mitten, what advice or suggestions do you have for your colleagues and future teachers?

GOING FURTHER

1. Read and discuss the Teacher Notes for "The Mitten Problem" probe (Keeley, Eberle, and Farrin 2005 pp. 104–108). Pay particular attention to the Related Research and Suggestions for Instruction and Assessment sections.

2. Two of the difficulties related to this probe are: (1) the difference in how we use the word *heat* in our everyday language versus the scientific use of the word and (2) misunderstanding the concept of temperature. Read and dis-

cuss the section on Core Idea PS3 Energy on pages 120–126 in *A Framework for K-12 Science Education* (NRC 2012) or online at *www.nap.edu/openbook. php?record_id=13165&page=120.*

3. Read and discuss the article Teaching for Conceptual Change (Watson and Konicek 1990). This article can be accessed through the Exploratorium's website at *www.exploratorium.edu/ifi/resources/teachingforconcept.html.*

4. Examine the NSDL Science Literacy Strand Map on Energy Transformation at *http://strandmaps.nsdl.org/?id=SMS-MAP-2071.* Discuss how the idea of heat builds progressively across K–12.

5. Read and discuss the teacher notes that accompany the everyday science mystery story, "Warm Clothes" mentioned in the chapter (Konicek-Moran 2010).

6. Watch and discuss the NSTA archived web seminar on the *NGSS* core idea of *Energy* at *http://learningcenter.nsta.org/products/symposia_seminars/NGSS/webseminar29. aspx.*

7. Try out the interactive Science Object on Energy: Thermal Energy, Heat, and Temperature in the NSTA Learning Center at: *http://learningcenter.nsta.org/product_detail. aspx?id=10.2505/7/SCB-EN.3.1.*

Chapter 8
Is It Living?

The word *living* is commonly used throughout elementary science lessons that focus on the biological world. It is a word teachers often take for granted when teaching life science concepts. When a teacher uses the word *living* when talking with students, or students encounter the word *living* when reading about science or gazing at posters on the wall, is their constructed meaning of the word *living* the same as the meaning intended by the teacher or instructional materials? How similar the constructed meaning of a common word like *living* is to the meaning intended by the teacher or instructional materials depends on how a student makes meaning out of the language teachers use so freely throughout their instruction. Formative assessment probes can be used to reveal the differing ways elementary students think about common words that make up the language of the science classroom.

Osborne and Freyberg (1985) point out the mismatch between familiar words used in the science classroom such as *living, animal, plant,* and *consumer* and the specialist meaning of these words. Often this mismatch goes undetected and passes as "understanding" when students use a word correctly in the science classroom. For example, students might say "an animal is a living thing" which is correct. However, does this mean the student has an accurate conception of what the word *living* means? The "Is It Living?" probe (Figure 8.1) can be used to uncover the hidden meanings children construct for the word *living* (Keeley, Eberle, and Farrin 2005).

Students' responses to this probe are not much different from the results of studies that were conducted in the 1980s around children's ideas about living and nonliving (Driver et al. 1994). The probe reveals the various attributes students use to decide whether something is living and how students interpret those attributes. For example, some students choose fire as a living thing. Their reason for choosing fire, along with boy, rabbit, and other living things on the list is because living things "eat." To the student, fire fits this rule because it needs to be "fed" wood or other fuel to keep burning. As the teacher listens to the student describe his or her reason for selecting fire, the teacher notes the mismatch between the common use of the word *eat* and the biological meaning of *eat.*

Some students believe living things need to be "active" and will select nonliving things like clouds, river, wind, and the Sun to fit this criterion. The pupae, seed, tree, and the hibernating bear may not fit the students' conception of the word *active* as it applies to living things. Active to them may mean visibly moving about.

Growth is a criterion for life but the word *grow* often has a different meaning to younger students. The common meaning of *grow* is "to get bigger" whereas a biological meaning of growth relates to the division and multiplication of cells in an organism. To younger students, clouds and fire *grow* because they get bigger.

Using the Probe

This probe is best used in a discussion context and is an effective instructional tool used to promote "science talk" and develop the norms of argumentation. Provide students with an opportunity to discuss their ideas in small groups first, and then engage students in a whole-class discussion. In one classroom I observed recently, third graders were discussing the probe in groups of four, deciding which things on the list were living. They worked together to then come up with a list of attributes they could use to decide whether something is considered living. These attributes and the examples they selected to fit these attributes were recorded in their science notebooks as their "beginning ideas." The teacher then asked each small group to select one student to sit in the *fishbowl,* a formative assessment classroom technique used to encourage careful listening and evidence-based arguments (Keeley 2008).

The five students in the fishbowl, representing each small group, sat in the center of the floor mat while the other students gathered around them. With their science notebooks in hand, the five students shared their thinking, challenging each others' ideas that differed from the small group they initially discussed their ideas in, while the other students listened carefully and made note of similar or differing ideas. The teacher carefully orchestrated the discussion in the fishbowl by asking probing questions and questioning (but not correcting) inconsistencies in their arguments.

After the fishbowl students finished their discussion, the other students had an opportunity to ask questions of the students in the fishbowl and provide feedback on the ideas that surfaced during the fishbowl discussion. The teacher then engaged the entire class in developing a set of ideas about "What Is Living" that represented the class's best thinking so far. Using the information she formatively gathered during the small-group, fishbowl, and whole-class discussions, she designed a series of learning stations that would confront

---

FIGURE 8.1.
The Is It Living? Probe

Is It Living?

Listed below are examples of living (which includes once-living) and nonliving things.
Put an X next to the things that could be considered living.

____ tree   ____ egg

____ rock   ____ bacteria

____ fire   ____ cell

____ boy   ____ molecule

____ wind   ____ Sun

____ rabbit   ____ mushroom

____ cloud   ____ potato

____ feather   ____ leaf

____ grass   ____ butterfly   ____ fossil   ____ mitochondria

____ seed   ____ pupae   ____ hibernating bear ____ river

Explain your thinking. What "rule" or reasoning did you use to decide if something
could be considered living?

---

students with their ideas and provide an opportunity for them to work out the inconsistencies in their ideas about living things as well as the scientific meaning of words they used to describe living things such as *grow, eat, move*, and *reproduce*.

It is also important to point out that when using a probe with elementary students, some students have no constructed meaning for some of the examples listed on the probe. Because these probes are intended to be used in grades K–12, words that are unfamiliar to students or developmentally inappropriate should be removed from the list. For example, the third-grade teacher using this probe removed *bacteria, cell, molecule*, and *mitochondria*. Although students may have heard some of these words, they lacked the scientific understanding at this early age to relate these examples to living or nonliving. It is also important to make sure all students are familiar with each word listed on or used with a probe, particularly with students who are English language learners. Teachers may want to include pictures with the words.

Taking the time to understand how children use common everyday words in the science classroom, and to help students understand that some words have a different meaning in science compared to their everyday use can promote deeper conceptual learning. Far too often we assume that children have the same meaning for common words that the teacher and the instructional materials intend. However, merely correcting students and providing them with definitions does not alleviate the language mismatch problem. This probe, as well as many others in the *Uncovering Student Ideas in Science* series can be used to help teachers plan instruction that connects language with learning.

REFERENCES

Driver, R. A. Squires, P. Rushworth, and V. Wood-Robinson. 1994. *Making sense of secondary science: Research into children's ideas.* London and New York: RoutledgeFalmer.

Keeley, P. 2008. *Science formative assessment: 75 strategies for linking assessment, instruction, and learning.* Thousand Oaks, CA: Corwin Press.

Keeley, P., F. Eberle, and L. Farrin. 2005. *Uncovering student ideas in science, volume 1: 25 formative assessment probes.* Arlington, VA: NSTA Press.

Osborne, R., and P. Freyberg. 1985. *Learning in science: The implications of children's science.* Portsmouth, NH: Heinemann.

INTERNET RESOURCE

Uncovering Student Ideas in Science series
www.nsta.org/publications/press/uncovering.aspx

NSTA CONNECTION

Read the introduction to *Uncovering Student Ideas in Science, Volume 1,* and download a full-size "Is It Living?" probe at *www.nsta.org/SC1104.*

Is It Living?
Reflection and Study Guide

QUESTIONS TO THINK ABOUT AFTER YOU READ THIS CHAPTER

1. Living is an example of a word teachers use that may have hidden meanings when used by children. What is meant by a hidden meaning in science? What are some other common words in science that may hold hidden meanings for your students?

2. After reading the chapter, what are some of the hidden meanings students have for characteristics that define life?

3. How would you use this probe in a discussion format? How can you foster productive science talk in your classroom?

4. How is the Fishbowl FACT (formative assessment classroom technique) used with this probe (Keeley 2008, 2014)? How would you use the fishbowl FACT with your students?

5. What are some things you would remove from this probe for your grade level? What are some things you would add?

6. Examine the different items on the justified list. Why do you think some of these items were included? What student ideas about living were they designed to elicit?

7. How could this probe be modified to uncover students' ideas about living, once living, or never living? How could you modify it for English Language Learners or emerging readers?

8. Many children's books anthropomorphize nonliving things such as putting a smiling face on a Sun. Can you find other examples in children's literature that anthropomorphize non-living things? How do you deal with anthropomorphism when using children's stories that have non-living characters?

9. Would you immediately correct students after hearing them use a common word incorrectly in a science context? Why or why not?

10. What does it mean to "argue" in science? How can you develop the scientific practice of argumentation with a probe such as this one? (For more information about the scientific and engineering practices of engaging in argument from evidence, visit NSTA's *NGSS* portal at: *http://ngss.nsta.org*.)

PUTTING FORMATIVE ASSESSMENT INTO PRACTICE

1. What did you learn about the characteristics your students use to determine whether something is living?

2. What are some of the basic needs your students recognize as common to living things? What are some of the basic life processes your students recognize as common to living things? Did your students apply any of these needs or processes to nonliving things? If so, how will you address their misunderstanding?

3. Which things on the list did children readily identify as living? Which things elicited misunderstandings that seemed to be common among your students?

4. A study by Stavy and Wax (1989) revealed that children seem to have different views for "animal life" and "plant life." In general, animals were more often recognized as living things. Did you find similar thinking among your students?

5. What were some of the common rules or reasoning your students used to decide whether something on the list was living?

6. What other information did you get from this probe that will inform your teaching of life science concepts?

7. Based on what you learned about your students' ideas about the concept of "living," what advice or suggestions do you have for your colleagues and future teachers?

GOING FURTHER

1. Read and discuss the Teacher Notes for the "Is It Living?" probe (Keeley, Eberle, and Farrin 2005, pp. 124–130). Pay particular attention to the Related Research and Suggestions for Instruction. There is quite an extensive body of research related to children's ideas about living things. Which of the instructional suggestions would you use in your teaching?

2. Understanding characteristics of living things is a precursor to ideas related to structure and function. Read and discuss the section on Core Idea LS1 From Molecules to Organisms: Structures and Processes on pages 143–150 in *A Framework for K–12 Science Education* (NRC 2012) or online at *www.nap.edu/openbook. php?record_id=13165&page=143.*

3. Read and discuss the article: Legaspi, B. and W. Straits. 2011. Living or Nonliving? First-grade lessons on life science and classification address misconceptions. *Science and Children* 38 (8): 27–31.

4. Watch and discuss the NSTA-archived web seminar on the *NGSS* core idea of From Molecules to Organisms: Structures and Processes at *http://learningcenter. nsta.org/products/symposia_seminars/NGSS/webseminar35.aspx*

With a Purpose

The first thing that comes to mind for many teachers when they think of assessment is testing, quizzes, performance tasks, and other summative forms used for grading purposes. Such assessment practices represent only a fraction of the kinds of assessment that occur on an ongoing basis in an effective science classroom. Formative assessment is a type of classroom assessment used by teachers to support instruction and learning (NRC 2001). It fits well into inquiry-based instruction because it is easily embedded into activities and rich classroom discussions. It enables teachers to make data-based decisions about next steps for learning while students are concurrently engaged in examining their own ideas and their peers' ideas. However, for formative assessment to be used effectively, it needs to be *purposefully* linked to learning and rooted in good teaching practices.

One of the questions I frequently get when I am working with schools to help teachers implement formative assessment practices is, "When do we use the probes?" The answer is that a formative assessment probe can be used at any point in an instructional cycle as long as there is a clear purpose. Regardless of the instructional model you use in your classroom, there are common stages in most that are identified in the Science Assessment, Instruction, and Learning (SAIL) Cycle represented in Figure 9.1, p. 63 (Keeley 2008).

Ainsworth and Viegut (2006, p. 109) stated "A good idea—poorly implemented—is a bad idea." One of the keys to successful implementation of an instructional resource, such as the formative assessment probes, is to be purposeful. Explicitly identifying your reason for using a probe and linking it to the stage of instruction at which its purpose is best achieved to support learning is a vital step in selecting an assessment probe. Using the stages in the SAIL Cycle, I will reflect upon and share some of the ways the probes you have encountered during the 2010–2011 issues of *Science and Children* can be used to purposefully integrate assessment, instruction, and learning.

Engagement and Readiness

In this combined stage, students' curiosity and interest in the content is activated. It is also the stage at which teachers determine students' readiness to learn the content. For example, the "Needs of Seeds" probe (Chapter 6) in the February 2011 issue of *Science and Children* can be used at the beginning of a unit to interest students in the content, activate their thinking, and generate curiosity to ask questions about seeds. At the same time the teacher gathers valuable data about students' prior knowledge and experiences related to living things and seeds. This assessment data helps the teacher determine prerequisite learning goals that may need to be addressed before students are ready to engage with the concepts and skills that make up the learning targets.

Eliciting Prior Knowledge

Elicitation is the process of drawing out students' existing ideas about a scientific process or phenomenon. During this stage the teacher identifies preconceptions the students bring to their learning and uses the information to design appropriate learning experiences that will challenge students' existing ideas to support conceptual change. At the same time it provides a metacognitive opportunity for students to surface their own ideas, become more aware of others' thinking, and identify what they know or need to know more about during the teaching and learning process.

For example the "Doing Science" probe (Chapter 1) in the September 2010 issue of *Science and Children* can be used before students engage in a full investigation to determine whether they have misconceptions related to how the "scientific method" is used. This data informs the approach the teacher takes to helping students understand there are a variety of ways scientists and students investigate phenomena without being locked into a rigid methodology.

The "Floating Logs" probe (Chapter 2) in the October 2010 issue of *Science and Children* reveals the intuitive rules that children use to reason about floating and sinking phenomena. For example, students may think that two different-size objects made of the same material float differently because larger objects sink more; an intuitive rule called "more–A, more–B." The probe provides an opportunity for students to examine their own ideas and develop explanations to support their thinking. At the same time, the teacher uses this valuable information to design a floating and sinking experience in which students will be confronted with their preconceptions and given the opportunity to construct alternative ways of thinking about the phenomenon.

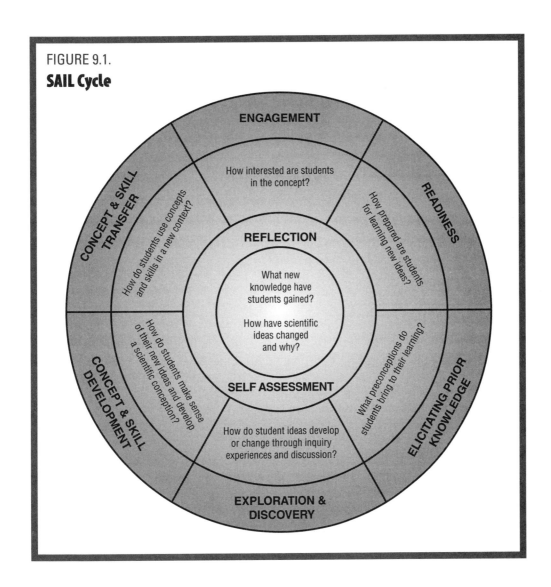

FIGURE 9.1.
SAIL Cycle

ENGAGEMENT

How interested are students
in the concept?

CONCEPT & SKILL
TRANSFER

How do students use concepts
and skills in a new context?

READINESS

How prepared are students
for learning new ideas?

REFLECTION

What new
knowledge have
students gained?

How have scientific
ideas changed
and why?

SELF ASSESSMENT

How do students make sense
of their new ideas and develop
a scientific conception?

CONCEPT & SKILL
DEVELOPMENT

What preconceptions do
students bring to their learning?

ELICITATING PRIOR
KNOWLEDGE

How do student ideas develop
or change through inquiry
experiences and discussion?

EXPLORATION &
DISCOVERY

Exploration and Discovery

In this stage students have an opportunity to explore and discover new ideas through rich scientific discussions or investigations. The teacher listens carefully for evidence of students' thinking as they engage in discussion or interact with materials. The probe can be used to initiate a prediction and explanation and then launch into inquiry. As students make observations, they sometimes see that their observations do not match their prediction, which leads to further investigation or discussion of alternative explanations. "The Mitten Problem" (Chapter 7) described in the March 2011 issue of *Science and Children* is an example of a probe to use before students begin an investigation into heat and temperature. The probe provides the context for launching into inquiry as well as valuable information about students' preconceptions. This information is taken into account as the teacher monitors the inquiry investigation and guides students toward the discovery that some objects themselves do not generate heat but rather slow down the transfer of heat.

Concept and Skill Development

After students have had an opportunity to engage in inquiry and small-group discussion, the teacher guides the class through the sense-making process that leads to the development of conceptual understanding of formal scientific concepts and skills. Assessment probes are used at this critical juncture to determine the extent to which students have grasped a concept or skill, recognized patterns or relationships among ideas, and connected their experiences to appropriate terminology. Assessment data identifies the need for further instructional experiences and readiness to move on to new concepts and skills.

For example, the assessment probe "How Far Did It Go?" (Chapter 5) could be used at this point in a series of lessons in which students describe motion and position. The teacher uses it to make sure students understand the importance of recognizing how to use a nonzero starting point to measure distance traveled. The probe provides an opportunity for teachers to decide whether to provide further opportunities to reinforce this important measurement skill before moving on to the next series of lessons. It is also an opportunity to provide feedback to the student or the class so that they can self-assess or recognize their measurement error and consider ways to correct it.

Concept and Skill Transfer

An important aspect of learning for understanding is being able to transfer ideas learned in one context to a new context. Formative assessment probes written in the form of a justified list are useful in analyzing how students make generalizations or connect "learned" ideas to examples not encountered previously. These assessment probes also provide a metacognitive opportunity for students to think about how to use their knowledge and skills in new situations or with different examples. The justified list probe "Does It Have a Life Cycle?" (Chapter 3) from the November 2010 issue of *Science and Children* can be used to determine how well students can transfer ideas about life cycles from the organisms they studied in their instructional unit to a variety of other organisms. The information often reveals the extent students were bound by the context in which they encountered the ideas and helps inform the teacher about adjustments that may need to be made to the instructional unit. It may also reveal the need for additional learning opportunities to ensure students can transfer ideas from one context to another. For example, after teaching a butterfly unit, the teacher may use this probe and discover that the students think only organisms similar to a butterfly or organisms that go through complete metamorphosis have life cycles. This is a signal to the teacher that the big idea that all organisms go through a life cycle (that may differ) needs to be further developed before concluding the instructional unit.

Self-Assessment and Reflection

These are critical aspects of any instructional cycle that happen throughout the learning process and provide opportunities for students to monitor their own thinking and learning. Reflection can also culminate in an opportunity for students to compare what they thought before or during a sequence of instruction and how they now make sense of the content. Using an assessment probe for self-assessment and reflection during the unit or at the end provides valuable feedback to the teacher in knowing how students perceive how their ideas have changed or deepened.

For example, the "Is It Living?" probe (Chapter 8) in the April/May 2011 issue of *Science and Children* can be given as a self-assessment to determine the degree of confidence students have in their own ideas. The probe asks students to check off the things they consider to be living. In addition, the teacher can ask them to circle the answers they are most confident about and underline the ones they are unsure about. This provides an opportunity for students to self-assess the extent of their own knowledge while indicating to the teacher the examples that are most problematic. As a reflection it can be returned to students at the end of an instructional sequence. They then reflect on how their ideas have changed from when they first responded to the probe. Students can be given a new copy of the probe in which they may change their initial responses and provide an explanation on how their thinking changed and why.

Reflection is probably the most overlooked aspect of assessment and instruction both in student and professional learning. As you conclude this school year, I encourage you to reflect on your formative assessment experiences. How has your knowledge of formative assessment increased? What are you doing differently as a result of reading these probes? What else do you need to know to use formative assessment probes effectively? What can you do over the summer to plan for the effective implementation of formative assessment when you start the new school year? Because reflection is part of a cycle, it doesn't end with these questions. Hopefully your responses will feed back into actions you will take to continue your quest to *purposefully* integrate assessment, instruction, and learning.

REFERENCES

Ainsworth, L., and D. Viegut. 2006. *Common formative assessments.* Thousand Oaks, CA: Corwin Press.

Keeley, P. 2008. *Science formative assessment: 75 practical strategies for linking assessment, instruction, and learning.* Thousand Oaks, CA: Corwin Press.

National Research Council (NRC). 2001. *Knowing what students know: The science and design of educational assessment.* Washington, DC: National Academies Press.

INTERNET RESOURCE

Uncovering Student Ideas in Science series
  www.nsta.org/publications/press/uncovering.aspx

NSTA CONNECTION

Read the introduction to *Uncovering Student Ideas in Science, Volume 1,* and download the probes referenced in this article, at *www.nsta.org/SC1107.*

● ●

With a Purpose
Reflection and Study Guide

QUESTIONS TO THINK ABOUT AFTER YOU READ THIS CHAPTER

1. What is formative assessment? How is it different from summative assessment?

2. Formative assessment and good instruction are like two sides of the same coin. What does that statement mean to you?

3. What does the quote, "A good idea poorly implemented, is a bad idea" mean to you (Ainsworth and Viegut 2006 p. 109)? What does this quote have to do with formative assessment?

4. What does it mean to be purposeful about formative assessment. How can using an instructional model or identifying stages in an instructional cycle help you be more purposeful about using formative assessment?

5. Examine Figure 9.1 and the descriptions of the SAIL Cycle in the chapter. Which of these stages of instruction correspond to the instructional model or method of instruction you use in your science teaching (e.g., the 5E model)? How does your use of formative assessment link to each of these stages?

6. How does the SAIL Cycle help you think about linking assessment, instruction, and learning?

7. This chapter provides examples of how the probes discussed in the previous chapters are used in various stages of an instructional cycle. Can you think of ways one of these probes can be used in a stage other than the one described? Think of other probes you have used in the *Uncovering Student Ideas in Science* series. In which stage of instruction did you use the probe?

8. Reflection and self-assessment are two important components of an instructional cycle that are often overlooked. Why are they important? How can you purposefully build in time for reflection and self-assessment when using a formative assessment probe?

PUTTING FORMATIVE ASSESSMENT INTO PRACTICE

1. What will you do differently after reading this chapter?

2. How will you modify your instructional materials to purposefully embed formative assessment and strengthen the link between assessment, instruction, and learning?

3. Based on what you learned from this chapter, what advice or suggestions do you have for your colleagues and future teachers?

GOING FURTHER

1. Read and discuss the Introduction to *Uncovering Student Ideas in Science, Volume 2: 25 More Formative Assessment Probes* (Keeley, Eberle, and Tugel 2007). This introductory chapter on pages 1–9 addresses the link between instruction and assessment. This chapter can be accessed free online at *www.nsta.org/store/product_detail. aspx?id=10.2505/9780873552738.*

2. Read and discuss Chapter 2 Integrating FACTs with Instruction and Learning in Keeley, P. 2008. *Science formative assessment: 75 practical strategies for linking assessment, instruction, and learning.* Thousand Oaks, CA: Corwin Press.

3. Learn more about purposeful instruction and assessment by reading and discussing the book: Tweed, A. 2009. *Designing effective instruction: What works in science classrooms.* Arlington, VA: NSTA Press.

4. Read and discuss the article: Volkman, M., and S. Abell. 2003. Seamless assessment. *Science and Children* 40 (8): 41–45.

Chapter 10

Where Are the Stars?

Gazing at the night sky is a familiar experience for many elementary students. Depending on where children live, they can often look out a window and see the Moon and stars (though in urban areas some children may not be able to observe stars in their immediate environment). Some students can even distinguish between "very bright stars" and planets. Children have "seen" the night sky in other ways as well. They may have seen the Moon and stars in television shows, movies, posters, or children's picture books. Regardless of whether they see the Moon and stars firsthand or vicariously, the Moon and the stars in the night sky are a familiar sight.

"Emmy's Moon and Stars" (Figure 10.1, p. 71) is a formative assessment probe designed to find out whether students recognize where the stars are in relation to the Earth and the Moon (Keeley, Eberle, and Tugel 2007). The Earth and the Moon are both part of our solar system. What about stars? Where do students think the stars are in relation to other bodies in our solar system? How far away do they think they are?

Sixty-four fifth graders field-tested this probe. Prior to fifth grade, the students in the field test learned about the solar system. Some students studied the solar system in third grade; others in fourth grade. From the sample of students who answered this probe after learning about the solar system, the results were surprising. When asked to describe where the Moon and the stars were that Emmy could see through her window, this is what they thought:

- 19% of the students chose the best answer—A. There are no stars between the Earth and the Moon.
- 15% chose answer B. One star is between the Earth and the Moon.
- 41% chose answer C. A few stars are between the Earth and the Moon.
- 10% chose answer D. There are many stars between the Earth and the Moon.
- 15% chose answer E. There are several stars between the Moon and the edge of our solar system.

In all, 19% of the students chose the correct response, whereas 81% of the students chose an incorrect response. What does this tell us about students' ideas and how we can use this data to inform instruction (which is the purpose of formative assessment)? To answer that question, we need to examine the students' thinking about the Earth, Moon, stars, and their position in the solar system and beyond. The selected responses alone do not provide much information about what each student is thinking.

For this reason, every assessment probe in the *Uncovering Student Ideas in Science* series is two-tiered. It begins with a selected response section in which students are given answer choices to pick from that best match their thinking. The second part of the probe is an open-response format in which students are asked to explain their thinking. Although a quick tally of selected answer choices can give the teacher a sense of the extent to which the class understands the concept being addressed, it is the careful examination of students' thinking in the second part that reveals the various ideas students drew upon to select their answer choice. This information can come from students' written explanations or from class discussions using a variety of FACTs (formative assessment classroom techniques) that engage students in defending their ideas (Keeley 2008).

Guiding Instruction

Sometimes students' ideas come from misinterpretations of taught concepts, misrepresentations in books and other media, pieces of scientific ideas that have not yet been strung together in a cohesive and coherent way, intuition, or their own "common sense" way of thinking about phenomena. As far as formative assessment goes, it is this second part of the probe that yields the most useful information about students' ideas teachers can use to inform their next steps. Whether planning for whole-class instruction or differentiation, formative assessment guides instructional decisions.

For example, several students who chose answer B explained that the Sun is a star and it is between the Earth and the Moon. The students are correct in recognizing the Sun is a star, but their explanation points out the need for further instruction that will help them understand the Sun is the only star in our solar system and the planets and Moons orbit around the Sun. Some students mentioned a small star between the Earth and the Moon because "that is what it looks like when I see the Moon and sometimes a star is close to it." I wonder whether some students' ideas were influenced by picture books? I have even seen drawings in which a star is nestled within the curve of a crescent Moon, which means it would be in front of the Moon in relation one's to view!

Students who chose B and D generally described small stars being "sprinkled" or spread out between the Earth and the Moon because they could see little stars "all

The Emmy's Moon and Stars Probe

Emmy's Moon and Stars

Emmy looked out her window and saw the Moon and stars. She wondered how far away they were. Circle the answer that best describes where you think the Moon and stars are that Emmy sees.

A There are no stars are between the Earth and the Moon.

B One star is between the Earth and the Moon.

C A few stars are between the Earth and the Moon.

D There are many stars between the Earth and the Moon.

E Several stars are between the Moon and the edge of our solar system.

Explain your thinking.

through the sky around the Moon." Although further probing would be necessary, this may point out the need for instruction that would help students realize many of those stars are larger than our Sun, and all are much larger than our Moon. Because they are so far away, they look like tiny points of light. Furthermore, children's picture books often make it look like the stars are "around the Moon." Response E revealed two interesting ideas: (1) the notion that the stars are all "behind the Moon" and (2) the notion that stars are spread out between the planets and the rest of the solar system. Although they did not use the word, some responses seemed to indicate the idea of a galaxy without recognizing the stars in the galaxy are far away, outside of the solar system. For example, one student explained, "There are millions and millions of stars surrounding planets and everything way out far in the solar system."

Examining the teacher notes for this probe reveals that even an interviewed high school student described stars as being dispersed within the realm of the solar system

(Agan 2004). The teacher notes also point out the difficulty students have dealing with large magnitudes. Numbers like billions and trillions are incomprehensible to most children and even adults. If you drew a line on a whiteboard representing a billion miles and asked a student (or adult!) to quickly point out where a million miles would be on the line, many children will point to the midpoint of the line. The number that represents a million miles would actually be near the beginning of the line.

The point is that for assessment to be formative, teachers need to probe beyond the selected response. Multiple-choice items can be good assessments if well-designed. They can be even better assessments, especially for formative use, if students are asked to share their thinking in either written or oral form. These "enhanced multiple-choice" assessments provide opportunities to probe much deeper and uncover ideas that may serve to inform significant curricular and instructional decisions. One simple change you can make to your own selected-response-item assessments is to ask students to explain why they chose an answer. That simple inquiry into students' thinking can provide you with a treasure trove of information to inform your teaching and promote learning. As a result of examining students' ideas about "Emmy's Moon and Stars," maybe you will reconsider the efficacy of the solar system model activities you have been using. Perhaps the assessment results will help you think about new ways to help students model and conceptualize the solar system within part of the larger universe.

REFERENCES

Agan, L. 2004. Stellar ideas: Exploring students' understanding of stars. *Astronomy Education Review* 3 (1): 77–97.

Keeley, P. 2008. *Science formative assessment: 75 practical strategies for linking assessment, instruction, and learning.* Thousand Oaks, CA: Corwin Press.

Keeley, P., F. Eberle, and J. Tugel. 2007. *Uncovering student ideas in science, volume 2: 25 more formative assessment probes.* Arlington, VA: NSTA Press.

INTERNET RESOURCE

Uncovering Student Ideas in Science series
www.nsta.org/publications/press/uncovering.aspx

NSTA CONNECTION

Read the introduction to *Uncovering Student Ideas in Science, Volume 1,* and download a full-size "Emmy's Moon and Stars" probe at *www.nsta.org/SC1109.*

Where Are the Stars? Reflection and Study Guide

QUESTIONS TO THINK ABOUT AFTER YOU READ THIS CHAPTER

1. Examine the fifth-grade student responses to the probe, "Emmy's Moon and Stars." Did the percentages of incorrect responses surprise you? How do you think your students would respond to the same probe?

2. Formative assessment probes have two parts: a selected response section where students select an answer that most matches their thinking and a constructed response section where students have to provide an explanation. How does a quick tally of the first part, the selected responses, give you a picture of your students' thinking? Does that part provide you with all the information you need to address your students' ideas? What additional information does the second part of a probe provide?

3. How can reading the Related Research summaries provided in the teacher notes to a probe help you anticipate ideas your students are likely to have?

4. How is an enhanced multiple-choice question different from a standard multiple-choice question? Which is better for formative purposes? How can you turn a standard multiple-choice question in your instructional materials into an enhanced multiple-choice question?

5. Why do you think it is difficult for students to understand vast distances in space? How does a child's concept of scale, size, and distance in the universe change as he or she progresses from one grade level span to the next?

6. Student responses to this probe may indicate the need to examine the solar system or night sky models you use to teach students about objects in our solar system and the night sky. What types of models do you currently use in your instruction? How does this chapter help you think differently about the models you use?

7. *A Framework for K–12 Science Education* and *The Next Generation Science Standards* emphasize scientific and engineering practices as one of the three dimensions of learning science. One of these practices is developing and using models. How does this probe support the use of this scientific practice? (For more information about scientific and engineering practices visit NSTA's *NGSS* portal at: *http://ngss. nsta.org*.)

PUTTING FORMATIVE ASSESSMENT INTO PRACTICE

1. What did you learn about your students' ideas by examining their responses to the probe? Were you surprised by any of their responses?

2. Compare the total number of students who chose incorrect responses to the number of students who chose the correct response. What does this data tell you?

3. If possible, compare student responses from different grade levels at least two years apart. How are the responses similar or different across grades?

4. What formative instructional decisions did (or will) you make as a result of examining your students' responses to this probe? How do the results of this probe indicate the need for modifications to your curriculum, instructional materials, or assessments?

5. How do the results from this probe inform the development of learning targets for your students? What would be an example of a learning target you would develop after examining your students' responses and what success criteria could you develop and share with students to provide evidence of understanding?

6. Based on what you learned from using this probe with your students, what suggestions do you have for your colleagues and future teachers?

GOING FURTHER

1. Read and discuss the Teacher Notes for the "Emmy's Moon and Stars" probe (Keeley, Eberle, and Tugel 2007, pp. 178–183). Pay particular attention to the Related Research and Suggestions for Instruction and Assessment sections.

2. Read and discuss the section on the crosscutting concept of Scale, Proportion, and Quantity on pages 89–91 in *A Framework for K–12 Science Education* (NRC 2012) or online at: *www.nap.edu/openbook.php?record_id=13165&page=89*. Discuss how the concept of scale can be used to help students comprehend what they see in the night sky.

3. The concept of stars and where they are located is included in the *NGSS* and *A Framework for K–12 Science Education* under Core Idea ESS1-Earth's Place in the Universe. Read pages 173–176 or online at *www.nap.edu/openbook.php?record_id=13165&page=145* and discuss what your students should be expected to know about objects they see in the night sky.

4. Children's books include representations of the solar system and the night sky. How could you use NSTA's annual list of Outstanding Science Tradebooks to help students learn about stars and objects in the visible night sky (See *www.nsta.org/publications/ostb*)?

5. There are several additional probes that target children's ideas about what they see in the night sky in the book, *Uncovering Student Ideas in Astronomy* (Keeley and Sneider 2012).

6. Read and discuss the guest editorial: Nelson, G. 2008. Building ladders to the stars. *Science and Children* 45 (1): 8.

7. Watch and discuss the NSTA-archived web seminar on the *NGSS* practice of developing and using models at *http://learningcenter.nsta.org/products/symposia_seminars/NGSS/webseminar6.aspx*.

8. Watch and discuss the NSTA-archived web seminar on the *NGSS* crosscutting concept of scale at: *http://learningcenter.nsta.org/products/symposia_seminars/NGSS/webseminar21.aspx*.

9. Watch and discuss the NSTA-archived web seminar on the *NGSS* core idea Earth's Place in the Universe at: *http://learningcenter.nsta.org/products/symposia_seminars/NGSS/webseminar31.aspx*.

Chapter 11
Pushes and Pulls

S tudents have early childhood experiences with basic force concepts well before they encounter the word *force* in the science classroom. For example, it doesn't take long for a child to figure out that pushing or pulling on a toy will cause it to move in the direction of the push or pull. From a very early age, children push and pull on the objects they play with. The forces that young children are most familiar with are ones exerted by their own or someone else's bodies. Their observations of forces usually involve human actions or actions by other living things such as a puppy pushing a ball away with its nose. As a result, many children develop an idea of forces being equated with living things and movement long before they learn about forces in the elementary science classroom. The way the word *force* is used in everyday language affects students' understanding of force in a science context. For example, young children often associate the word *force* with coercion, physical activity, and muscular strength (Driver et al. 1994). This transfers to their notion of a push or pull as well.

When the concept of force is first taught in the elementary curriculum, it is usually introduced as a push or a pull. *A Framework for K–12 Science Education* describes grade-band endpoints for the Core Idea: Motion and Stability: Forces and Interactions (NRC 2012). It states that by the end of grade 2 students should know that objects pull or push each other when they collide or are connected.

What do we know about students' ideas related to pushes and pulls? Early research studies have shown that some students have difficulty associating manifestations of force with pushes or pulls. For example, some primary-age children do not associate a kick or a throw with a push. Furthermore, some students believe there is a force that just holds things in place without pushing or pulling, such as a book on a table (Driver et al. 1994; Minstrell 1982).

Students' preconceptions about pushes or a pulls may affect their learning about types of forces and the interactions involved. The friendly talk probe "Talking About Forces"

FIGURE 11.1.
The Talking About Forces Probe

Talking About Forces

Five friends were talking about forces. This is what they said:

Rae: "I think a push is a force and a pull is something else."

Scott: "I think a pull is a force and a push is something else."

Yolanda: "I think a force is either a push or a pull."

Miles: "I think forces are neither pushes nor pulls. I think they are something else."

Violet: "I think pushes and pulls are forces, but there is also another type of force that just holds things in place."

Which friend do you agree with the most? _____

Explain your thinking. Describe what you think a force is.

(Figure 11.1) can be used to uncover students' preconceptions of the familiar words *push* and *pull* and what they mean in the context of forces (Keeley and Harrington 2010). A friendly talk assessment probe is a type of selected response assessment probe in which students select the character they agree with the most (Keeley 2008). With younger students (K–2) it is best to adapt the probe using only the first three answer choices: Rae, Scott, and Yolanda. The information gathered by the teacher can be used to inform learning opportunities in which primary age students can explore and identify a variety of pushes and pulls.

For example, if some students identify force with a push but not a pull, that is an indication to the teacher that students need opportunities to experience both pushes and pulls as forces. A teacher might take students on a "Push and Pull Walk" to identify examples of things they observe being pushed or pulled. Or, a teacher can use a card sort activity in which students are given a set of pictures or descriptions of an action printed on cards that they then sort into examples of forces that are pushes, forces that are pulls, and ones they are not sure about. Listening to the students share their reasons why something is an example of a push or a pull provides insights into students' thinking about forces and interactions between objects.

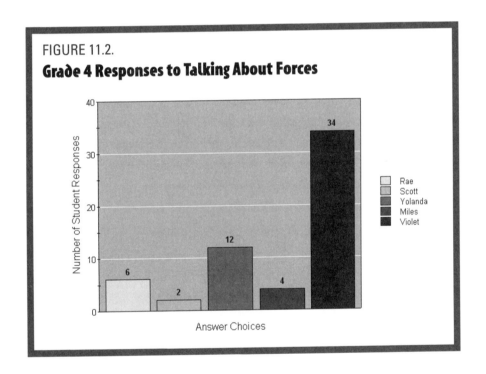

FIGURE 11.2.

Grade 4 Responses to Talking About Forces

Years later, after students have been taught basic ideas about forces as pushes and pulls, students may still hold on to their own interpretations of pushes and pulls. For this reason, it is helpful to use formative assessment probes to uncover students' ideas about important prerequisite concepts "learned" in prior grades, before building toward progressively more complex concepts. Figure 11.2 shows results from 58 fourth graders who responded to this probe prior to beginning a unit on force and motion. The students had previously encountered the basic concept of force as a push or pull in second grade.

As you can see from the graph, most of the students selected Violet, which matches the research finding that a "holding force" is a commonly held idea. The idea of a holding force may originate well before students reveal this common misconception in middle school or the physics classroom. Most of the students who agreed with Violet generally explained that pushes and pulls are forces because they make things move, but there is another type of force that keeps things from moving. Several students used the example of a refrigerator magnet "having a force that stuck the magnet to the refrigerator." Students who selected Yolanda typically explained that both pushes and pulls cause movement so they are examples of forces. These students equate force with movement.

Overall, the assessment probe reveals that students need opportunities to analyze forces where motion is obvious, less obvious, as well as forces acting on objects at rest. *A Framework for K–12 Science Education* states that by the end of fifth grade, students should know that "An object at rest typically has multiple forces acting on it, but they counterbalance one another."

Although there is more research to be done on diverse students' ideas related to force and motion, particularly with younger children's notion of a push or pull, teachers can gather and analyze their own assessment data to understand their students' ideas and make informed decisions about appropriate learning pathways. The new *Framework for K–12 Science Education* has provided us with updated descriptions of the content and sequence of learning expected of all students. Using formative assessment probes, such as the example provided, can help teachers understand where their own students' ideas are in relation to the Core Ideas described in the *Framework*. Furthermore, the probes can reveal conceptual difficulties your own students may face as they build toward progressively more detailed and sophisticated understandings.

Connecting formative assessment probes to the Core Ideas described in *A Framework* will help you move toward the vision of science teaching and learning that both the *Framework* and the *Uncovering Student Ideas in Science* series embodies. Now that the *Framework* has been released, let's continue to explore this powerful connection between formative assessment and the core ideas. And remember, the probes are not formative unless you use the results to inform your teaching and promote learning!

REFERENCES

Driver, R., A. Squires, P. Rushworth, and V. Wood-Robinson. 1994. *Making sense of secondary science: Research into children's ideas.* London and New York: RoutledgeFalmer.

Keeley, P. 2008. *Science formative assessment: 75 practical strategies for linking assessment, instruction, and teaching.* Thousand Oaks, CA: Corwin Press.

Keeley, P., and R. Harrington. 2010. *Uncovering student ideas in physical science, volume 1: 45 new force and motion assessment probes.* Arlington, VA: NSTA Press.

Minstrell, J. 1982. Explaining the "at rest" condition of an object. *The Physics Teacher* 20: 10–14.

National Research Council (NRC). 2011. *A framework for K–12 science education: Practices, cross-cutting concepts, and core ideas.* Washington, DC: National Academies Press.

INTERNET RESOURCE

Uncovering Student Ideas in Science series
   www.nsta.org/publications/press/uncovering.aspx

NSTA CONNECTION

Download a full-size "Talking About Forces" probe at *www.nsta.org/SC1110*.

• •

Pushes and Pulls
Reflection and Study Guide

QUESTIONS TO THINK ABOUT AFTER YOU READ THIS CHAPTER

1. Children have experiences with forces before they start school. They also encounter the word *force* in their everyday language. How do their everyday experiences outside of school and language affect they way they think about forces in science class?

2. Students initially learn about forces through pushes and pulls. How do students' preconceptions about pushes and pulls affect their thinking about forces?

3. "Talking About Forces" is an example of a probe format frequently used in the *Uncovering Student Ideas in Science* series, called a friendly talk probe. How is this format different from a typical multiple-choice question? What do you like about this format?

4. Notice how the picture in this probe and the names used for the answer choices portray diversity. Why is it important for students in the science classroom to see examples of people representing diverse groups talking about their ideas?

5. Examine each of the answer choices to the probe. What kinds of instructional experiences would you provide to address each of these answer choices?

6. How do you think student responses in grades K–2 may differ from students in grades 3–5? How do you think students' responses in grades 3–5 may differ from students in grades 6–8? What do you think may account for these differences?

Are there some ideas about pushes and pulls that may be resistant to change across grade spans?

7. If pushes and pulls are taught in the K–2 curriculum, why would a teacher in grades 3–5 use this probe? Are formative assessment probes limited to the grade level in which a specific idea is taught?

8. Examine the data from Figure 11.2. Grade 4 Responses to Talking About Forces. What do the data tell you? What factual statements can you make from the data? What inferences might you draw from the data?

9. How is it useful for teachers to collaboratively examine student thinking by pooling their data and having data-driven discussions about student thinking? How can you use collaborative inquiry into examining student thinking in your school or other setting?

10. How is it useful to examine and reference readings from *A Framework for K–12 Science Education* (NRC 2012), the *Benchmarks for Science Literacy* (AAAS 2009), or the *Atlas of Science Literacy* (AAAS 2001, 2007) when analyzing student data from a formative assessment probe?

PUTTING FORMATIVE ASSESSMENT INTO PRACTICE

1. What did you learn about your students' understanding of forces as pushes and pulls? Were you surprised by any of their answer choices and explanations?

2. What ideas did your students use to support their answer choices? Do their responses reveal evidence of the source of their misconception or their understanding?

3. What modifications will you make to your curriculum, instruction or instructional materials based on students' responses to this probe?

4. Did you engage in collaborative discussions about the student work from this probe with other teachers who used the same probe? What protocol or process did your group use to examine and discuss the data? What kinds of decisions did your group come up with based on the discussion?

5. Based on what you learned about your students' ideas about pushes and pulls, what advice or suggestions do you have for your colleagues and future teachers?

GOING FURTHER

1. Read and discuss the Teacher Notes for the "Talking About Forces" probe (Keeley and Harrington 2010, pp. 72–74). Pay particular attention to the Related Research and Suggestions for Instruction and Assessment sections.

2. Read and discuss the section on Core Idea PS2 Motion and Stability: Forces and Interactions on pages 113–120 in *A Framework for K–12 Science Education* (NRC 2012) or online at *www.nap.edu/openbook.php?record_id=13165&page=113*

3. Examine the NSDL Science Literacy Strand Map on Laws of Motion at: *http://strandmaps.nsdl.org/?id=SMS-MAP-1357*. Follow the strand "forces and motion" and discuss how the idea of pushes and pulls begins in K–2 and build progressively.

4. Examine and discuss the *Benchmarks for Science Literacy* (AAAS 2009) research summary on forces. Scroll down to 4F Motion: The Concept of Force. *www.project2061.org/publications/bsl/online/index.php?chapter=15&section=C&band=4#4*.

5. Watch and discuss the NSTA archived web seminar on the *NGSS* core idea of Motion and Stability: Forces and Interactions at *http://learningcenter.nsta.org/products/symposia_seminars/NGSS/webseminar28.aspx*.

Chapter 12
Teachers as Classroom Researchers

"The practice of teaching, even for teachers who do not think of themselves as researchers, is one in which the elements of research are present. Asking questions about practice, collecting evidence, making sense of the evidence, and sharing conclusions with others: These activities are happening all over schools during every school day, and they are elements of research" (Roberts, Bove, and van Zee 2007, p. ix).

Elementary teachers have a treasure trove of data about students' ideas at their fingertips every day. In 1999 a group of researchers from Indiana University and Purdue University decided to tap into this wealth of data about students' ideas in science by inviting NSTA members to participate in a research study about children's conceptions of animals. Published in *Science and Children* (Barman et al. 1999), the article received an overwhelming response from K–8 teachers across the nation interested in participating in this national study by collecting and submitting their own student data about students' perceptions of animals. A year later, results of that study were published in *Science and Children* (Barman et al. 2000).

The "Is It an Animal?" justified list probe (Figure 12.1) is similar to the set of examples used in Barman's study. It includes a variety of animals as well as a few nonanimal examples. About half of the choices are the same as the ones used in Barman's study, the other half are different. This was intentional in order to see whether children's answers differed if an alternative example within the same class of organisms was used. In the probe, students are also encouraged to describe the "rule" or reasoning they used to decide whether something on the list is an animal. For younger and ELL students, picture cards can be used with a sorting strategy in place of the text-based probe. As students sort each of the picture cards, they describe why they think each example is or is not an animal (picture cards available online; see NSTA Connection).

This probe is particularly useful to the elementary teacher as the concept of "animal" appears frequently in the elementary curriculum, children's stories and trade books, and popular television programs. Children are naturally interested in animals. However, does their conception of what an animal is match the scientific description of an animal? What criteria do children use to determine if an organism is an animal, and what informs the criteria they use? These are interesting questions to pursue in the classroom. Although I have seen hundreds of student responses to this probe and have had teachers share their findings with me, I'm not going to share what I have learned in this column! Instead, I'm encouraging you to consider conducting your own study with your own students and perhaps engaging other colleagues in your school to collaborate with you.

An Action QUEST

Using a systematic approach to pursue questions about students' thinking and collecting, analyzing, and sharing these data is a form of informal teacher classroom research, often described as action research. Action research deepens teachers' understanding of teaching and learning and provides an opportunity for teachers to contribute to the professional knowledge base by sharing findings with a colleague, with the entire school, or even through published articles in *Science and Children*. Formative assessment probes can lead to new discoveries about student learning in science. A natural way to systematically use the probes to informally study students' ideas in your own classroom is through engaging in a QUEST Cycle (Rose Tobey and Minton 2011). The QUEST Cycle was developed by my mathematics colleague, Cheryl Rose Tobey, for the *Uncovering Student Thinking in Mathematics* series and includes the following steps (adapted for use in science):

Questioning student understanding of a science concept
Uncovering student understanding using a formative assessment probe
Examining student responses to the assessment probe
Seeking links to cognitive research and other studies
Teaching implications: what will I do as a result of my findings?

Q: Start with a question. What do you want to learn more about in terms of students' ideas related to the concept of an animal? Your question might be: *What kinds of organisms do my students classify as animals? What criteria do they use to decide if an organism is an animal?*

U: The next step is to identify a formative assessment probe that can help you answer the question you posed. You could use the "Is It an Animal?" probe (Figure 12.1; Keeley, Eberle, and Farrin 2005). After selecting the probe, you need to decide how you will administer it. Will you use it as a paper/pencil assessment and collect students' written responses? Will you use it in small-group and whole-class discussions, recording students' ideas as they share them verbally with the class? Will you interview individual

FIGURE 12.1.

The Is It an Animal? Probe

Is It an Animal?

Which of the organisms listed are animals? Put an X next to each organism that is considered to be an animal.

___ cow   ___ spider

___ tree   ___ snail

___ mushroom   ___ flower

___ human   ___ monkey

___ worm   ___ beetle

___ tiger   ___ whale

___ shark   ___ frog   ___ mold

___ starfish   ___ chicken   ___ snake

Explain your thinking. Describe the "rule" or reasoning you used to decide if something is an animal.

students? Once you decide on the probe and a data collection technique, you are ready to dig into a gold mine of student data right in your own classroom!

E: After collecting the data, it is time to examine students' thinking. Tally the student responses to the answer choices to get a classroom picture of which organisms students chose as animals and which ones they chose as nonanimals. You might decide to organize your data in a table similar to the one from Barman's study. Next, you will examine the students' explanations for why something is considered to be an animal or not an animal. This part of the probe reveals some of the commonly held ideas students have about animals and the overgeneralizations they tend to make. Categorizing their ideas and keeping track of similar misconceptions reveals common patterns of understanding/misunderstanding that help explain why they chose which examples on the formative assessment probe are considered animals.

S: Now that you have a picture of your own students' ideas, it's time to seek out what others have learned through their research. Using the "Related Research" section in the teacher notes that accompany each formative assessment probe in the *Uncovering Student Ideas in Science* series, you can compare your own students' ideas to findings that have come from formal research studies. You might even look at some of the suggested articles in the NSTA journals that describe studies of students' ideas about animals, such as the Barman articles (NSTA members can access these articles online in the *Science and Children* archives). Do your findings confirm what has been described in the research? Perhaps you have discovered new insights into children's thinking that extend the findings in the articles cited in the research.

T: Being aware of your own students' ideas and findings from the research literature are important, but your QUEST doesn't end there. For assessment to be formative, you need to use findings from both your classroom and formal research studies to consider implications for your teaching of the concept. The teacher notes for each probe include Curricular and Instructional Considerations by grade level as well as a section on Suggestions for Instruction and Assessment. Read these over carefully, noting suggestions that you might use with your students to address the ideas you uncovered in your classroom. Perhaps there are changes you need to make to the curriculum or specific instructional strategies that may be useful. There is no one-size-fits-all solution to addressing student learning. The instructional decisions you make will depend on what you find out about your own students and their ideas about animals.

Finally, this is not a linear process. It often cycles back to asking a new question. For example, you might find that the majority of your students do not select humans as an example of an animal. This is further supported by Barman's study. This might lead to a new question that you want to explore further, such as *Do Students Consider Humans to Be Animals?* In this case, you might use the "No Animals Allowed" probe (Keeley 2011; see NSTA Connection), specifically designed to address the common children's conception that humans are not animals, and go through the same cycle again.

Whether you use "Is It an Animal?" or any of the other probes in the *Uncovering Student Ideas in Science* series, I think you will see how naturally these formative assessment probes encourage a deep curiosity about students' ideas and easily lead to informal research in your classroom. As Roberts, Bove, and van Zee (2007) discovered when working with teacher researchers, "once one experiences 'doing' teacher research, one cannot let go"!

REFERENCES

Barman, C., N. Barman, K. Berglund, and M. J. Goldston. 1999. Assessing students' ideas about animals. *Science and Children* 37(1): 44–49.

Barman, C., N. Barman, M. Cox, K. Berglund-Newhouse, and M. J. Goldston. 2000. Students' ideas about animals: Results of a national study. *Science and Children* 38 (1): 42–47.

Keeley, P., F. Eberle, and L. Farrin. 2005. *Uncovering student ideas in science, volume 1: 25 formative assessment probes.* Arlington, VA: NSTA Press.

Keeley, P. 2011. *Uncovering student ideas in life science: 25 new formative assessment probes.* Arlington, VA: NSTA Press.

Roberts, D., C. Bove, and E. van Zee. 2007. *Teacher research: Stories of learning and growing.* Arlington, VA: NSTA Press.

Rose Tobey, C., and M. Minton. 2011. *Uncovering student thinking in mathematics: 25 formative assessment probes for the elementary classroom.* Thousand Oaks, CA: Corwin Press.

INTERNET RESOURCE

Uncovering Student Ideas in Science series
www.nsta.org/publications/press/uncovering.aspx

NSTA CONNECTION

Download a full-size "Is It an Animal?" probe at *www.nsta.org/SC1111*.

● ●

Teachers as Classroom Researchers Reflection and Study Guide

QUESTIONS TO THINK ABOUT AFTER YOU READ THIS CHAPTER

1. How is the practice of teaching one in which "the elements of research are present"? What do you do as a teacher that is similar to what researchers do?

2. What does it mean to "have a treasure trove of data at your fingertips every day"? How can the formative assessment probes help you mine this data?

3. Some probes in the *Uncovering Student Ideas in Science* series are considered precursor probes. A precursor probe is one which uncovers students' ideas about basic concepts that are included in the standards but are not included as a standard by itself. For example there are several probes that target ideas related to matter (conservation of matter, properties of matter, states of matter) but there is no standard that targets the basic concept of matter. How is the probe, "Is It an Animal?" considered a precursor probe for elementary science? How is it related to the standards in your curriculum?

4. What is action research? Have you ever conducted action research? How can the probes be used to conduct action research in your classroom?

5. How might you use the QUEST Cycle with a formative assessment probe and the teacher notes that accompany a probe to conduct classroom research?

6. How does teacher action research mirror some of the scientific and engineering practices in *A Framework for K–12 Science Education* (NRC 2012) and the *NGSS*? (For more information about scientific and engineering practices visit NSTA's *NGSS* portal at *http://ngss.nsta.org*.)

PUTTING FORMATIVE ASSESSMENT INTO PRACTICE

1. What did you learn about your students' understanding of animals after using this probe? Were you surprised by any of their answer choices and explanations?

2. What "rule" or reasoning did your students use to support their answer choices? Do their responses reveal evidence of the source of their misconceptions or their understanding?

3. What modifications will you make to your curriculum, instruction or instructional materials based on students' responses to this probe?

4. What could you do next to design and carry out an Action Research Project? What would your question be? What probe would you use to conduct your research?

5. Based on what you learned about your students' ideas about animals, what advice or suggestions do you have for your colleagues and future teachers?

6. Based on what you now know about the value of conducting action research with the probes, what suggestions do you have for your colleagues in using action research for professional development?

GOING FURTHER

1. Read and discuss the Teacher Notes for the "Is It an Animal" probe on pages 118–122 (Keeley, Eberle, and Farrin 2005). Pay particular attention to the Related Research and Suggestions for Instruction and Assessment sections.

2. A similar version of this probe is designed for primary-grade children (Keeley 2013). Examine the K–2 version of "Is It an Animal?" and read and discuss the Teacher Notes on pages 10–13.

3. Read and discuss the article on the Barman study at: Barman, C., N. Barman, K. Berglund, and M. Goldston. 1999. Assessing students' ideas about animals. *Science and Children* 37 (1): 44–49. Follow up by reading and discussing Barman, C., N. Barman, M. Cox, K. Berglund-Newhouse, and M. J. Goldston. 2000. Students' ideas about animals: Results of a national study. *Science and Children* 38 (1): 42–47.

4. Read and discuss an article that shows how action research was used with the "Wet Jeans" probe: Tugel, J. and I. Porter. 2010. Uncovering student thinking in science through CTS action research. *Science Scope* 33 (1): 30–36.

5. Learn more about teacher research by reading the NSTA Press book, *Teacher Research: Stories of Learning and Growing* (Roberts, Bove, and van Zee 2007).

Chapter 13

Representing Microscopic Life

In 1683 Anton Van Leeuwenhoek wrote a letter to Britain's Royal Society describing the "animalcules" he observed under the lenses he designed as one of the first early microscopes. He described them as if they were tiny animals, because details of their internal cell structure were unknown at that time. Through his writing, Leeuwenhoek expressed the joy and excitement of seeing the diversity of this newly discovered microscopic life.

When children in the intermediate grades (grades 3–5) first look through a microscope at a drop of water from a pond, they experience the same excitement in discovering this teeming world of microscopic life that Leeuwenhoek described. The typical microscopes children use reveal a wide assortment of fascinating shapes and sizes of various algae, protozoans, and small multicell animals such as rotifers and their variety of intricate motions as they zip and glide through the water. However, the limitations of microscopes used in elementary school usually fail to reveal details of the internal structure of these single-cell and tiny multicellular organisms. As students observe these organisms, some will make generalizations about internal structures they cannot see. These generalizations may remain hidden and follow these students from one grade to the next if they never surface. Formative assessment probes and probing techniques can be used to uncover generalizations students make as they grapple with new concepts and unfamiliar observations.

When observing microscopic life, the elementary learning intention should be on recognizing that "some living things consist of a single cell. Like familiar organisms, they need food, water, and air; a way to dispose of waste; and an environment they can live in" (AAAS 2009). Details of their internal structures (organelles) can wait until later grades. However, teachers can use this opportunity to uncover students' preconceptions about the internal organization of single-cell organisms and link their ideas to what they know about the structure of animals. How do children compare the structure of these tiny one-cell organisms to familiar animals? What do they think they would see if they had a microscope powerful enough to look inside these tiny one-cell organisms? The formative assessment probe, "Pond Water," (Figure 13.1, p. 94) reveals how elementary children will often apply what they know about animal structures to these newly discovered organisms, connecting their knowledge of the familiar to the unfamiliar through overgeneralization.

Do Students Overgeneralize?

The formative assessment probe, "Pond Water," is designed to find out whether students use their knowledge of animal structure to overgeneralize about the internal structure of single-cell pond organisms (Keeley 2011). Namely, do they think the organisms they observe in a drop of water have organs similar to humans and other familiar animals? Research on learning indicates that when children encounter a new concept, they strive to connect the new concept to something they already know. Prior to discovering single-

FIGURE 13.1.

The Pond Water Probe

Pond Water

Six students were looking at a drop of pond water through a microscope. They were amazed to see many different types of tiny single-celled organisms moving around in the drop of water. The students wondered what they would see if they had a more powerful microscope. They wondered how the insides of the single-celled organisms compared to the insides of animals. This is what they said:

Lanny: "I don't think they have any of the organs animals have."

Dorothy: "I think they have a few of the organs most animals have."

Seamus: "I think they have a few of the organs simple animals, like worms, have."

Nick: "I think the only organs they have that animals also have are the digestive organs."

Valynda: "I think the only organs they have that animals also have are the organs they use for breathing."

Brian: "I think they have all of the organs that most animals have, they are just a lot smaller."

Which person do you agree with the most? _____ Explain why you agree.

Describe what you think you would see if you could look inside a single-celled organism with a powerful microscope.

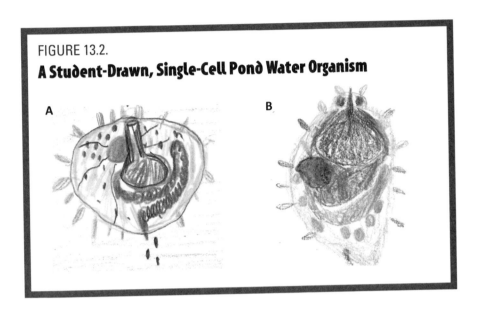

FIGURE 13.2.
A Student-Drawn, Single-Cell Pond Water Organism

A   B

cell life, students have observed and learned about a variety of animals. Instruction was focused at the macro level as students explored the external and internal structures that enable humans and other animals to carry out their life processes and survive. They learned about a variety of body organs, such as the heart, stomach, and lungs, and body systems such as the digestive system. Do students overgeneralize by applying what they learned about animal structure to single-cell organisms?

A study by Dreyfus and Jungwirth (1989) conducted with 16-year-old Israeli students revealed confusion about levels of organization in living things. Even though they had been taught about cells in previous years, several of the students thought that single-cell organisms contained organs such as hearts, lungs, and stomachs. Results from the "Pond Water" probe showed similar ideas held by fifth-grade students. The most common response among 43 fourth- and fifth-grade students who participated in a field test of this friendly talk probe was the answer choice, "I think they have a few of the organs most animals have." Their explanations included reference to these organisms having basic organs such as eyes, brains, hearts, lungs, and stomachs.

Furthermore, when some students were asked to draw what they thought the insides of a single-cell organism might look like, more details emerged that revealed how students connect their prior knowledge to these new, unfamiliar organisms. Figure 13.2A and B are examples of drawings made by fifth graders in response to the question, "If you had a powerful microscope to look inside a single-cell pond water organism, what do you think the insides would look like? Draw what you think you would see."

When the two students who drew Figure 13.2A and B were asked to describe their drawings, both students pointed out the stomach and heart in their drawings. In addition, the student who drew Figure A pointed out the intestines. Figure B was described as having lungs and a head with eyes. When asked why they thought the pond water organisms had organs like animals have, both students similarly replied that they needed organs and a digestive and circulatory system to live, breathe, and eat.

Although these drawings were made prior to responding to the probe and were dependent on having students explain their drawings for the teacher to interpret them accurately, probes such as "Pond Water" could be extended in a similar way by combining the probe with the formative assessment classroom technique (FACT) *annotated student drawings* (Keeley 2008). *Annotated student drawings* are student-made, labeled illustrations that visually represent and describe students' thinking about a scientific concept. They are not drawings made by direct observation of an object or organism. Rather they are representations of the mental image students have in their heads. Annotating the drawing with labels and descriptions provides further information that teachers can formatively use. Annotated drawings allow the teacher to see and the student to reveal understandings or misunderstandings that often remain hidden from written response procedures (White and Gunstone 1992). With younger students or English language learners, teachers can employ informal interview techniques and annotate the students' drawings as children explain their ideas to the teacher. The drawing provides a visual means of explanation and justification of the answer choice that makes up the second part of every assessment probe. As in the "Pond Water" example, drawings often reveal generalizations students make based on their prior learning.

Moving Learning Forward

Once this information is revealed, teachers can think about how to best use these incorrect generalizations to move students' learning forward in a grade-level appropriate way. Although it is not productive for elementary students to learn names and details about cell organelles, this is an opportunity to point out to students that single-cell organisms are structurally different from the animals they learned about. They do not have organs such as hearts, lungs, and stomachs. Details will come in later grades when students learn about cells and cell structure.

As you use the formative assessment probes in the *Uncovering Student Ideas in Science* series, think about ways you can encourage children to draw their thinking. Visit the Uncovering Student Ideas website (see Internet Resource) to learn more techniques for using representations for formative assessment. Remember, "If a picture is worth a thousand words, perhaps drawing and visualizing can help science students enhance their learning potential" (NSTA 2006, p. 20).

REFERENCES

American Association for the Advancement of Science (AAAS). 2009. *Benchmarks for science literacy online. www.project2061.org/publications/bsl/online*

Dreyfus, A., and E. Jungwirth. 1989. The pupil and the living cell: A taxonomy of dysfunctional ideas about an abstract idea. *Journal of Biological Education* 23 (1): 49–53.

Keeley, P. 2008. *Science formative assessment: 75 practical strategies for linking assessment, instruction, and learning.* Thousand Oaks, CA: Corwin Press.

Keeley, P. 2011. *Uncovering student ideas in life science: 25 new formative assessment probes.* Arlington, VA: NSTA Press.

NSTA. 2006. Picturing to learn makes science visual. *NSTA Reports* 18 (2): 20.

White, R., and R. Gunstone. 1992. *Probing understanding.* Philadelphia, PA: Falmer Press.

INTERNET RESOURCE

Uncovering Student Ideas in Science series
    www.uncoveringstudentideas.org

NSTA CONNECTION

Download a full-size "Pond Water" probe at *www.nsta.org/SC1112*.

● ●

Representing Microscopic Life Reflection and Study Guide

QUESTIONS TO THINK ABOUT AFTER YOU READ THIS CHAPTER

1. Did you experience the same feeling of awe and excitement the first time you saw a drop of water teeming with life? What about the first time your students ever observed microscopic life? How was their reaction similar to Leeuwenhoek's?

2. Why do you think children tend to compare the internal structures of microscopic organisms to familiar animals?

3. What does it mean to "overgeneralize?" How can assessment probes such as "Pond Water" reveal the tendency to overgeneralize?

4. What is the advantage of asking students to draw a picture to explain their thinking? How could you combine a probe with the formative assessment classroom technique (FACT), Annotated Drawing (Keeley 2008, 2014)?

5. What instructional decision might a teacher make after examining the drawings in Figure 2?

6. Drawings are a type of model students use to represent and explain their thinking. *A Framework for K–12 Science Education* (NRC 2012) and the *NGSS* include the scientific and engineering practice of developing and using models, which includes drawings as representations of ideas. How can you use drawings with probes to support this practice? (For more information about scientific and engineering practices visit NSTA's *NGSS* portal at *http://ngss.nsta.org*.)

PUTTING FORMATIVE ASSESSMENT INTO PRACTICE

1. What did you learn about your students' ideas about the internal structure of microscopic life after using this probe? Were you surprised by any of their answer choices and explanations?

2. Did you ask your students to draw a picture to explain their thinking? What did you learn from your students' drawings that provided additional evidence of their ideas related to structure?

3. What modifications will you make to your curriculum, instruction or instructional materials based on students' responses to this probe?

4. Based on what you learned about your students' ideas about microscopic life, what advice or suggestions do you have for your colleagues and future teachers?

GOING FURTHER

1. Read and discuss the Teacher Notes for the "Pond Life" probe on pages 118–122 (Keeley, Eberle, and Farrin 2005). Pay particular attention to the Related Research and Suggestions for Instruction and Assessment sections.

2. Examine the matrix for Practice 2: Developing and Using Models on page 5 in the *NGSS* Appendix F to see how drawings are included by grade level (*NGSS* Lead States 2013). You can access Appendix F online at *http://23.23.182.104/access-standards*.

3. Watch and discuss the NSTA-archived web seminar on the *NGSS* practice of developing and using models at *http://learningcenter.nsta.org/products/symposia_seminars/NGSS/webseminar6.aspx*.

Chapter 14
The Daytime Moon

The familiar adage "seeing is believing" implies that children will recall a particular phenomenon if they had the experience of seeing it with their own eyes. If this were true, then most children would believe that you could see the Moon in both daytime and at night. However, when children are asked, "Can you see the Moon in the daytime?" many will say "no," even though they have actually seen the Moon many times in the morning or afternoon sky. The formative assessment probe, "Objects in the Sky," (Figure 14.1, p. 101) shows how persistent the belief is among elementary-age children that the Moon can only be seen in the nighttime (Keeley, Eberle, and Tugel 2007).

Learning Goals

Understanding where the Moon is located at different times of the day and its changing appearance as viewed from Earth are important goals for learning. In the early elementary grades, the *Benchmarks for Science Literacy* state that by the end of second grade, students should know that "the Sun can be seen only in the daytime, but the Moon can be seen sometimes at night and sometimes during the day. The Sun, Moon, and stars all appear to move slowly across the sky" (AAAS 2009). *A Framework for K–12 Science Education* (NRC 2012) states that by the end of second grade, students should understand that "Patterns of the motion of the Sun, Moon, and stars in the sky can be observed, described, and predicted." Building on this earlier idea, by the end of fifth grade, students should understand, "The orbits of Earth around the Sun and of the Moon around Earth, together with the rotation of Earth about an axis between its North and South poles, cause observable patterns. These include day and night; daily and seasonal changes in the length and direction of shadows; phases of the Moon; and different positions of the Sun, Moon, and stars at different times of the day, month, and year."

To achieve an understanding of these important learning goals, elementary students should have the opportunity to observe the position and phases of the Moon in both the daytime and nighttime sky and discover the cyclic pattern of Moon phases by analyzing their recorded observations. This experience is one of several critical prerequisites to constructing an explanation for the phases of the Moon. Before students engage in monthly observations to discover the pattern of Moon phases, consider using a probe such as "Objects in the Sky" to find out whether students recognize that the Moon can be seen in the daytime.

With Their Own Eyes

Parts of this probe are based on research conducted on children's ideas about the Moon. Vosniadou and Brewer (1994) found that many young children believe that the Moon is only visible at night and the occurrence of the Moon in the sky is associated with nighttime. Some students will even attribute the appearance of the Moon as a causal factor for night. Research has also revealed that some students believe the Moon rises straight up in the evening, stays at the top of the sky throughout the night, and then sets straight down (Plummer 2009). When children are asked why they think the Moon is only visible at night, they often explain their thinking using this "up–down rule" and may even confuse it with the rising and setting of the Sun.

Do these ideas change with age? Consider using the "Objects in the Sky" probe or asking the question, "When do you see the Moon: daytime, nighttime, or both?" across multiple grades from first- through fifth-grade. Share your data and look for differences in students' ideas. Probe further to find out why students think the Moon is only visible during the evening. If students believe the Moon can be seen in the daytime, probe to find out what phases they think can be observed during the day. Consider other factors that may have influenced their thinking that the Moon is visible only in the nighttime. For example, children's storybooks, trade books, and instructional materials almost always show the phases of the Moon in a dark, nighttime sky. Instead of telling students that the Moon is visible in the daytime, take them outside during a phase when it is visible in the morning or afternoon sky and let them see it for themselves. Continue making daytime observations of the Moon, when visible, to reinforce that we can sometimes see the Moon during the daytime. Have students record the time and position of the Moon in the daytime sky and the phase present.

This probe points out that when students have seen something, they don't necessarily believe it. By confronting students with their idea that the Moon is only visible at night through a direct experience of observing the Moon during the daytime, you may help your students give up their misconception. Furthermore, the probe is formative in nature by pointing out the importance of not limiting students' observations to a nighttime context. When the Moon is visible during the school day, encourage students to make their monthly observations then, as well as in the evening (which is necessary for them to see the full Moon because it rises when the Sun sets and sets as the Sun rises). By using this probe, perhaps you will see the old adage change to "believing is seeing!"

FIGURE 14.1.

The Objects in the Sky Probe

Objects in the Sky

Different things can be seen in the sky.

Put a **D** next to the things that are seen **only** in the daylight.

Put an **N** next to the things that can be seen **only** at night.

Put a **B** next to the things that can be seen in **both** day and night.

___ the Sun

___ the Moon

___ the next-nearest star to our Sun

___ constellations

Explain your thinking. How did you decide when you could see different things in the sky?

REFERENCES

American Association for the Advancement of Science (AAAS). 2009. *Benchmarks for science literacy. www.project2061.org/publications/bsl/online*

Keeley, P., F. Eberle, and J. Tugel. 2007. *Uncovering student ideas in science, volume 2: 25 more formative assessment probes.* Arlington, VA: NSTA Press.

National Research Council (NRC). 2012. *A framework for K–12 science education: Practices, crosscutting concepts, and core ideas.* Washington, DC: National Academies Press.

Plummer, J. 2009. Early elementary students' development of astronomy concepts in the planetarium. *Journal of Research in Science Teaching* 46 (2): 192–209.

Vosniadou, S., and W. Brewer. 1994. Mental models of the day/night cycle. *Cognitive Science* 18: 123–183.

INTERNET RESOURCE

Uncovering Student Ideas in Science series
www.uncoveringstudentideas.org

NSTA CONNECTION

Download the "Objects in the Sky" probe at *www.nsta.org/SC1201.*

• •
The Daytime Moon
Reflection and Study Guide

QUESTIONS TO THINK ABOUT AFTER YOU READ THIS CHAPTER

1. Why do you think many students believe the Moon is only visible at night, even though they have seen the Moon during the day?

2. What have researchers learned about children's ideas related to the Moon and when it is visible? How does this probe mirror findings from research into children's ideas?

3. Do your students keep a Moon journal or make observations of the Moon to determine the monthly pattern of Moon phases? Why is it a good idea to not limit their observations to only the night sky?

4. Can you think of other examples, besides seeing the Moon in the daytime, when the familiar adage "seeing is believing" does not hold true?

5. What if you added planets, such as Venus, to the answer choices. Do you think your students have seen a planet in the sky in the morning or before it gets dark? If they have seen a planet after sunrise or shortly before sunset, how do you think they would respond to the probe?

PUTTING FORMATIVE ASSESSMENT INTO PRACTICE

1. What did you learn about your students' ideas about seeing objects in the sky? Were you surprised by any of their answer choices and explanations?

2. What kinds of evidence did your students mention in their explanation? What does this tell you about how experience shapes their thinking?

3. What modifications will you make to your curriculum, learning targets, instruction, or instructional materials based on your students' responses to this probe?

4. Examine your instructional materials and children's books that include representations or photographs of the Moon. How many of the instructional resources you use show the Moon in the daytime?

5. If you had an opportunity to compare student responses to this probe across different grade levels, how did their responses compare? Does the frequency of the misconception that the Moon can only be seen at night change with grade level?

6. Based on what you learned about your students' ideas about seeing objects in the sky, what advice or suggestions do you have for your colleagues and future teachers?

GOING FURTHER

1. Read and discuss the Teacher Notes for the "Objects in the Sky" probe on pages 186–190 (Keeley, Eberle, and Tugel 2007). Pay particular attention to the Related Research and Suggestions for Instruction and Assessment sections.

2. Objects in the sky and their motions is included in the *NGSS* and *A Framework for K–12 Science Education* under Core Idea ESS1-Earth's Place in the Universe. Read pages 173–176 or online at *www.nap.edu/openbook.php?record_id=13165&page=145* and discuss how this probe relates to what students should know about objects they see in the sky.

3. Read and discuss an article that describes how young children investigated the Moon, including the discovery that the Moon is visible in the daytime. Fitzsimmons, P., D. Leddy, L. Johnson, S. Biggam, and S. Locke. 2013. The moon challenge. *Science and Children* 50 (1): 36–41.

Chapter 15
Seeing the Light

The elementary grades are pivotal years for developing interest in and positive feelings toward science and science learning. During the elementary years, children are active seekers, creators, and users of their own knowledge about the natural world. Good teachers provide opportunities for students to satisfy their curiosity through carefully designed activities that hone students' practices of observation for the purpose of discovering new knowledge about the way their world works. The ways children learn science and the confidence they build in developing, testing, making meaning of, and sharing their ideas can create a strong interest in science that carries over into the middle and high school years. That is, as long as students are not asked to set aside their own ideas and ways of testing them to simply accept what they are told or read in a textbook without any challenge to their preconceptions.

Formative assessment probes are valuable tools for gathering information about students' thinking for the purpose of informing instruction. However, there is a secondary outcome of using formative assessment probes that affects students' attitudes toward and interest in science while simultaneously contributing to their learning. To consider this outcome of promoting learning and interest in science through formative assessment probes, let's examine scenarios from two different fifth-grade classrooms where the teacher used the "Can It Reflect Light?" (Keeley, Eberle, and Farrin 2005; Figure 15.1, p. 107) probe to gather information about students' preconceptions related to reflection of light.

Classroom #1

Mr. Dokas is planning a series of lessons on light and how it travels. The day before he begins his lesson plan, he seeks to learn more about his students' ideas related to light reflection with the "Can It Reflect Light?" probe. He has each student silently complete it, making sure they include an explanation that describes the "rule" or reasoning they used to decide which objects can reflect light.

That evening, Mr. Dokas looks over the students' responses and notices that students chose only certain objects. Their explanations included ideas such as "things need to be shiny to reflect light" or "light colored things reflect light, dark things absorb it." He decides he will start the lesson the next day by first having students read the section in their textbook on light and light reflection, hoping it would dispel some of their misconceptions about reflection. He notices that his instructional materials included an activity using mirrors that would show students how light is reflected. He makes a note to make sure that after students complete this activity, he will be sure to emphasize that light also reflects off other objects, not just mirrors.

The next day Mr. Dokas starts the lesson by posting the learning target *To understand how light reflects off objects.* He tells his students that he read over their responses to the probe and noticed that most of the class chose objects such as water, mirror, shiny metal, clouds, and so on. He shared several of their explanations pointing out common reasons such as "things had to be shiny or light-colored." He tells the class to read pages 56–59 and summarize what they learned about light reflection. After discussing the reading, the students begin an activity on mirrors in which they measure the angle of incidence and angle of reflection to show how light is reflected from a mirror. Following the activity, Mr. Dokas puts the word *scattering* on the word wall and explains how scattering is light reflection from objects that are not smooth like a mirror. He then directs them back to the probe to decide if they think other objects on the list can reflect light with equal angles of incidence and reflection like a mirror or reflect light through scattering. He is pleased to see that most of the students now select most of the objects on the list. He notes that he will address some of the objects they are still struggling with when they get to the textbook section on light and color and do the other activities that are included in their instructional materials.

Classroom #2

Ms. Wilson is also planning a series of lessons on light and how it travels. She also begins her instructional planning by selecting the "Can It Reflect Light?" probe to uncover her students' preconceptions. She prints the objects on cards, creating enough sets for groups of three students. The next day she posts her learning intention for the afternoon's lesson, *To investigate how objects reflect light.* She points out the learning intention and begins the lesson by explaining the card sort activity. She reminds students of the norms for engaging in "science talk." Students work in small groups to sort their packet of cards into the things they think can reflect light, the things they think cannot reflect light, and the things they are unsure of or cannot agree on, justifying their reasons for each one.

Ms. Wilson circulates from group to group, observing how students sort the cards and listening carefully as they engage in respectful arguments to defend their thinking. She makes note of some of the reasons students use to justify their ideas about reflection. She is careful not to interject herself into the discussions or acknowledge whether their

FIGURE 15.1.

The Can It Reflect Light? Probe

Can It Reflect Light?

What types of objects or materials can reflect light? Put an X next to the things you think can reflect light.

___ water

___ gray rock

___ leaf

___ mirror   ___ dull metal

___ glass   ___ red apple

___ sand   ___ rough cardboard

___ potato skin   ___ the Moon   ___ milk

___ wax paper   ___ rusty nail   ___ bedsheet

___ tomato soup   ___ clouds   ___ brand new penny

___ crumpled paper   ___ soil   ___ old tarnished penny

___ shiny metal   ___ wood   ___ smooth sheet of aluminum foil

Explain your thinking. Describe the "rule" or the reasoning you used to decide if something can reflect light.

_____

_____

_____

ideas are right or wrong. At this point she wants her students to surface their own thinking first while she thinks about how to best move their learning forward based on the formative assessment data she is collecting through careful listening and observation.

She notices the students are highly engaged in discussion with each other. After everyone has had time to surface their ideas in small groups, she pulls the class together for a whole-group discussion. As a class they discuss which objects they are all sure can reflect light and which ones they don't agree on. She lists the "best thinking" so far from the class as to how to determine which objects can or cannot reflect light. She makes a note to herself that several of their ideas seem to match the research summary in the probe teacher notes, such as objects need to be shiny, smooth, or light colored. She also notes which objects, such as water, were informed by everyday experiences. Most of the students had been to the beach and explained how they got a sunburn from the reflection.

After Ms. Wilson posted a list of class ideas, she asked her students how they could find out whether the objects on the list could reflect light. She knew she had to provide a firsthand experience to the students in order for them to give up their strongly held misconceptions. As she looked through her instructional materials, she noticed they focused primarily on mirrors for teaching about reflection and later addressed the idea of absorption and reflection to understand why we see colors. She saw nothing in her materials that would develop the generalization that all objects we see reflect light. She realized the topic of light is familiar to students but also mysterious as they typically do not see light reflecting off non-shiny objects. She decided to let her students brainstorm how they could test the objects. One of the students explained how he played a game where he reflected light off mirrors with flashlights to make signals. He wondered if they could do a similar thing with flashlights and the objects on the list.

Ms. Wilson used that opportunity to guide them further toward an activity she had in mind and asked "What if we take Todd's idea and we darken the room, come up close to the shiny whiteboard, and hold our flashlights so they point away from the whiteboard toward the object. If light reflects off the object, what would we see on the whiteboard?" The students agreed they would see a spot of light. Some said they would even see the color on the whiteboard. Ms. Wilson further probed to ask them why this observation would provide the evidence they needed to accept that an ordinary object can reflect light. Almost all the students were in agreement that this was the evidence they needed. The next day she provided examples of the objects on the list, students came up to the whiteboard with different objects, the room was darkened and the testing began.

As students began to see the spot of light, and even colors, reflected on to the white-board, there was a buzz of excitement in the room. Ms. Wilson smiled as she heard shouts of "Come see this gray rock" or "Wow! Even this dull penny reflects light!" Soon the students were asking to test other objects in the room and the entire class was deeply engaged in "seeing the light."

After the activity, Ms. Wilson returned to the card sort activity and asked students which objects they would now sort differently. Cries erupted of "they all reflect light because we could see their reflection on the whiteboard." However, several students were at a loss to explain why. Ms. Wilson noted this for the next day's lesson where she would connect the role of light to how we see an object. But for now, she was pleased with her students' readiness to give up the idea that only shiny or light-color objects reflect light.

In returning to the learning intention, she asked her students to come up with success criteria to determine how well they met the learning intention for that day: *To investigate how objects reflect light.* They came up with (1) We can design a test to see if things can reflect light, (2) We can make a list of things that can reflect light, and (3) We can explain the evidence for how we know if something reflects light. She asked for thumbs up, thumbs sideways, and thumbs down for feedback on how well they thought they met the success criteria for the lesson. She was pleased to see that almost all students had thumbs up with a few thumbs sideways for the last success criteria.

Before ending the lesson, she posted the learning intention for the next day: *To understand how we see objects.* She told her students that tomorrow they would further examine the evidence for light reflection. Tonight she would begin planning a lesson to connect the activity they engaged in to how light reflects off objects and enters our eye. Now that her students could "see the light," her next step was to help them understand how "light helps us see."

Assessing *for* Learning

Consider the difference between the two classrooms. Both teachers were using formative assessment probes along with the important practice of making the intention of the day's lesson explicit to students. However, in the first classroom, the teacher neglected the opportunity for students to engage in exploration of their own ideas related to the probe through discussion and investigation. Instead, the teacher and the textbook were the "authorities" that dispensed knowledge to students. Furthermore, the activities did not directly address the students' ideas and were likely not to be very effective in addressing students' misconceptions about light. Instead of revising the lesson plan, Mr. Dokas made an attempt to explicitly address the misconceptions by giving students the information. However, he provided little or no firsthand opportunity for students to construct an understanding or engage in sharing their own thinking with the class. Instead they launched into an activity that only served to verify what they read in their textbook or were told by the teacher.

On the other hand, Ms. Wilson used the probe as an "assessment for learning." Her students engaged in social construction of ideas and evidence-based justification. She used her knowledge of students' ideas and the importance of giving them an opportunity to think about how they could test their ideas to launch into an exciting firsthand investigation that would allow them to see for themselves that all the objects on the list could reflect light. She regarded the student as the "meaning-maker." She made sure students saw the probe again and reflected on how their ideas had changed. She ended the lesson by revisiting the learning intention and success criteria so students could self-monitor and provide feedback on their own learning before she planned the next day's lesson.

As you compare these two scenarios of formative assessment probe use, I hope it "brings to light" the importance of creating an engaging learning environment where children have an opportunity to investigate the ideas they struggle with and construct new meaning through the personal participation of students in their own learning. The illuminating insights both teachers and students gain when using formative assessment probes to promote learning are sure to shine a new light into your classroom!

REFERENCE

Keeley, P., F. Eberle, and L. Farrin. 2005. *Uncovering student ideas in science, volume 1: 25 formative assessment probes.* Arlington, VA: NSTA Press.

INTERNET RESOURCE

Uncovering Student Ideas in Science series
    www.uncoveringstudentideas.org

NSTA CONNECTION

Download the "Can It Reflect Light?" probe at *www.nsta.org/SC1202*. Read the entire chapter and the introduction to *Uncovering Student Ideas in Science, Volume 1: 25 Formative Assessment Probes* in the NSTA Science Store (*http://bit.ly/vaihtZ*).

● ●
Seeing the Light
Reflection and Study Guide

QUESTIONS TO THINK ABOUT AFTER YOU READ THIS CHAPTER

1. Why do you think the elementary grades are considered pivotal years for developing positive feelings toward science and science learning?

2. Formative assessment is typically used to make informed decisions about teaching and learning. However, another outcome of using formative assessment is that it fosters learning while the teacher is collecting information about student thinking.

How do you think the use of a formative assessment probe, such as "Can It Reflect Light?" can promote learning as well as provide information about learning?

3. Examine the formative assessment probe, "Can It Reflect Light?" How did you respond (or would you have responded) to this probe prior to reading this chapter? What preconception(s) influenced your thinking related to this probe?

4. How would you describe Mr. Dokas's teaching? How would you describe Ms. Wilson's teaching? What constructive feedback would you give each teacher?

5. What are the differences between the two teachers and their style of teaching? In which classroom does formative assessment play a more central role? Which teacher would you say your teaching is most similar to, Mr. Dokas or Ms. Wilson?

6. What is meant by "social construction of ideas?" How can the use of a formative assessment probe help students think through and construct new ideas as they interact with their peers?

7. What role did self monitoring and feedback play in Ms. Wilson's classroom? How did Ms. Wilson use learning intentions and success criteria to move students' learning forward?

8. The *NGSS* Grade 4 performance expectation for PS4-Waves and Their Applications in Technologies for Information Transfer is "Develop a model to describe that light reflecting from objects and entering the eye allows objects to be seen" (NGSS Lead States 2013, p. 37). How does this probe prepare students for this performance expectation? What else are students expected to know and be able to do?

PUTTING FORMATIVE ASSESSMENT INTO PRACTICE

1. What did you learn about your students' ideas related to reflection? Were you surprised by any of their answer choices and the rule or reasoning used to explain their thinking?

2. What evidence of a common misconception did you find in your students' responses? Where do you think this misconception may have come from?

3. How is reflection of light typically taught in your science curriculum? What modifications will you make to your curriculum, learning targets, instruction, or instructional materials based on your students' responses to this probe?

4. Did you provide an opportunity for your students to discuss and share their ideas with each other as they responded to the probe? Did you use an interactive strategy such as a card sort? If so, what did you observe about your students as they discussed their ideas with their peers?

5. What post assessment could you (or did you) give your students to determine the extent to which they now understand reflection and the role of light in how one sees objects? What opportunity did you provide for your students to reflect on how their ideas may have changed?

6. Based on what you learned about your students' ideas about light reflection, what advice or suggestions do you have for your colleagues and future teachers?

7. What advice or suggestions do you have for your colleagues or future teachers about using formative assessment probes in small-group or whole-class discussions?

GOING FURTHER

1. Read and discuss the Teacher Notes for the "Can It Reflect Light?" probe on pages 26–30 (Keeley, Eberle, and Farrin 2005). Pay particular attention to the Related Research and Suggestions for Instruction and Assessment sections.

2. Reflection of light and the role of light in explaining how we see objects is included in the *NGSS* and *A Framework for K–12 Science Education* as core idea PS4.B: Electromagnetic Radiation. Read pages 133–136 or online at *www.nap.edu/openbook.php?record_id=13165&page=133* and discuss how this probe relates to what K–8 students should know about the reflection of light.

3. Read about and discuss other formative assessment classroom techniques (FACTs) that can be used with the probes in formats in which students share and discuss their ideas. A total collection of 125 FACTs are included in the *Science Formative Assessment* series (Keeley 2008, 2014).

4. Read and discuss the article: Matkins, J., and J. McDonnough. 2004. Circus of light. *Science and Children* 41 (5): 50–54.

5. Read and discuss The Early Years column: Ashbrook, P. 2012. Shining light on misconceptions. *Science and Children* 49 (2): 30–31.

Chapter 16
Food For Plants: A Bridging Concept

Do plants need food? This seems to be a simple question for students at all grade levels. For example, from grade 4 through high school most students recognize that unlike animals, plants are able to make their own food. However, when fifth-grade students were asked if plants needed food, some surprising ideas were revealed:

Student A: *"Yes, because they need to give food to animals that eat plants."*

Student B: *"No, they make food but they don't eat food like we do."*

Student C: *"No, because they have no mouth or teeth to eat and chew the food they make."*

Student D: *"Yes, they need plant food to grow." [When probed further, plant food comes from a jar one buys in a garden store.]*

Each of these responses shows parts of correct ideas that these fifth-grade students drew upon from their everyday experience and prior learning that they coupled with a misconception. For example, Student A understands the idea that animals depend on plants for food. However, this student seems to think that plants exist to make food for animals that eat plants. Student B knows that plants can make their own food, but the student thinks of food from a human perspective as something that gets "eaten." Student C recognizes that plants have structures different from plant-eating animals and that plants make food. However, this student thinks of food as something that is taken in through the mouth of an organism. Student D understands the important idea that like all organisms, plants need food to grow. However, when probed further to find out what the student meant by "plant food," a common misconception surfaces that plant food is a product purchased in a store that is given to plants as food to help them grow.

The student responses shown above match the research conducted over many years on children's commonly held ideas related to plants, food, and photosynthesis. For example, researchers have found that some students perceive the ability of plants to make their own food (photosynthesis) as something plants do for the benefit of people and animals (Roth and Anderson 1985). Research has also shown that students often lack the prerequisite scientific concept of food to build an understanding of photosynthesis as a process that provides energy and material for growth and repair for plants. Food is often perceived in an animalistic way as something that is taken into the mouth, chewed, and then goes to the stomach to be digested (Driver et al. 1994). Another commonly held idea described in numerous research studies is that plants take their food in through the soil, such as the "plant food" that is sprinkled on the soil. Using terminology like "plant food" to refer to fertilizer further contributes to this misconception (Driver et al. 1994).

Probing Misconceptions Further

Combined with the notion of a plant's need for food, the formative assessment probe "Is It Food for Plants?" (Keeley, Eberle, and Tugel 2007; Figure 16.1), further reveals the disconnect between the biological concept of food and students' ideas about what plants use as food. Although this probe is designed primarily for middle and high school students, it can be modified for grades 3–5 by substituting the answer choice *sugar* with *the food a plant makes*; substituting *carbon dioxide* and *oxygen* with *gases*, and removing *chlorophyll*. In this justified list probe, students are asked to check off the things on the list they think plants use as food, followed by an explanation of how they decided whether something is considered to be food for a plant. This probe can also be used as a card sort by putting the answer choices on cards and having students work in small groups to sort the cards into "things that are food for plants," "things that are not food for plants," and "things we do not agree on or are unsure about" (Keeley 2008). As students sort the cards, they must provide a justification as to why they consider something to be food or not food for plants.

As you listen to students engage in argumentation or analyze students' written responses to the probe, a clear picture emerges of the "common sense ideas" students hold. In order for anything to be considered food for an organism, it must be able to supply energy to the organism and provide the molecules an organism needs for growth and repair of structures. Therefore, the only thing on this probe's list that can be considered food for a plant is sugar, the carbohydrate a plant makes as a result of photosynthesis. Unlike animals, plants do not take food in directly from the environment. Instead, plants are like a "food factory," manufacturing their own food (sugar) from raw materials (carbon dioxide and water) using energy from sunlight to power the process. None of the other things on the list meet the biological criteria for food. However, they do meet a child's common sense view of food as things a plant takes in from its environment or needs to live.

In examining typical curricula and states' standards, students in the elementary grades typically learn that animals need air, water, and food, and plants need air, water, nutrients, and sunlight. In middle school they are introduced to the general process of

---

FIGURE 16.1.

The Is It Food for Plants? Probe

Is It Food for Plants?

Organisms, including plants, need food to survive. Put an X next to the things you think plants use as food.

___ sunlight

___ plant food from a garden store

___ sugar

___ carbon dioxide

___ minerals

___ fertilizer

___ soil

___ water

___ leaves

___ oxygen

___ chlorophyll

___ vitamins

Explain your thinking. How did you decide if something on the list is food for plants?

_____

_____

_____

---

photosynthesis, linked to plant structures. They learn that plants take in water through their roots and carbon dioxide through their leaves and use energy from sunlight captured by their leaves to make their own food. But do students really know what that food is and why plants need to make food? Furthermore, what is their concept of food and how might it impair students' ability to understand ideas related to plant needs at both a basic and increasingly sophisticated level?

Formative assessment probes often reveal the need to teach a bridging concept that may not be clearly articulated in state standards or instructional materials. An example of this is in the K–4 standard "Organisms have basic needs. For example, animals need air, water, and food; plants require air, water, nutrients, and light" (NRC 1996, p. 129) may unintentionally imply that animals need food but plants do not. When students get to middle school, the idea in the National Science Education Standards reads "Plants and some microorganisms are producers—they make their own food. All animals, including humans, are consumers, which obtain food by eating other organisms. Decomposers, primarily bacteria and fungi, are consumers that use waste materials and dead organisms for food" (NRC 1996, p. 158). Clearly there is no bridge between the elementary idea and the middle school idea that helps students understand what food is in a scientific sense and that plants also need food (not just make it).

Analyzing Probe Results

Examining students' responses from the "Is It Food for Plants?" probe reveals the need to address what food is in a scientific sense at an appropriate developmental level for elementary students and to further develop this concept at the middle level. For example, at the elementary level, students should know that all organisms use food and they get their food in different ways. Animals have to acquire it from their environment, but plants have the ability to make their own food. For example, the sugar we eat can be traced back to the sugar made by plants that we harvest from the sugar cane they make, or the maple syrup we eat on our pancakes comes from the sweet sap made by a maple tree. These phenomena provide explanatory evidence that plants make food in the form of sugars without going into the chemical details of photosynthesis. When students get to middle school, they are ready to combine ideas about energy and atoms to understand that the food plants make in the form of sugar, by transforming carbon dioxide and water with energy provided by sunlight, is used to carry out their life processes and that it also provides the building blocks (atoms and molecules) plants need to grow and produce new structures. It can be used immediately by the plant or stored for later use.

Each of these ideas at the elementary and middle school levels involves the bridging concept of food. Many of the probes in the *Uncovering Student Ideas in Science* series reveal the need to address bridging concepts. The *Framework* (NRC 2012) refers to this bridging of gaps as the "continuum of learning." The formative use of a probe can significantly improve curriculum and instruction by helping teachers see that the standards they are required to teach do not always stand alone. By uncovering students' ideas using formative assessment probes, teachers can identify the need to bridge the gap between and among some grade-level standards in order for students to fully achieve an in-depth understanding of core scientific ideas.

REFERENCES

Driver, R., A. Squires, P. Rushworth, and V. Wood-Robinson. 1994. *Making sense of secondary science: Research into children's ideas.* London: RouteledgeFalmer.

Keeley, P. 2008. *Science formative assessment: 75 practical strategies for linking assessment, instruction, and learning.* Thousand Oaks, CA: Corwin Press.

Keeley, P., F. Eberle, and J. Tugel. 2007. *Uncovering student ideas in science, volume 2: 25 more formative assessment probes.* Arlington, VA: NSTA Press.

National Research Council (NRC). 1996. *National science education standards.* Washington, DC: National Academies Press.

National Research Council (NRC). 2012. *A framework for K–12 science education: Practices, crosscutting concepts, and core ideas.* Washington, DC: National Academy Press.

Roth, K., and C. Anderson. 1985. *The power plant: Teachers' guide.* East Lansing, MI: Institute for Research on Teaching.

INTERNET RESOURCES

Uncovering Student Ideas in Science series
    www.uncoveringstudentideas.org

NSTA CONNECTION

Download the "Is It Food for Plants?" probe at *www.nsta.org/SC1204.*

Read the entire chapter and the introduction to *Uncovering Student Ideas in Science, Volume 1: 25 Formative Assessment Probes* in the NSTA Science Store (*http://bit.ly/vaihtZ*).

● ●
Food for Plants: A Bridging Concept
Reflection and Study Guide

QUESTIONS TO THINK ABOUT AFTER YOU READ THIS CHAPTER

1. How would you have answered the probe, "Is It Food for Plants?" prior to reading this chapter? What ideas about food or photosynthesis did you use as you thought about the best answer to this probe?

2. What does having "parts of correct ideas" mean when responding to a probe? What partially correct ideas did students A, B, C, and D have related to the question, "Do plants need food?" How did these ideas affect their thinking about food for plants?

3. Summarize some of the research related to this probe. How does this probe elicit some of these common misconceptions?

4. Biologically, how is food defined? How is the biological definition of food different from our everyday definition of food?

5. Would you use this probe as written with your grade level or would you modify the answer choices? If the latter, what modifications would you make for your grade level?

6. What should your students be expected to know at their grade level about the difference between how animals and plants get their food? What ideas will they learn at later grades that build upon these ideas?

7. What is a bridging concept and why are bridging concepts important? How is "food" considered a bridging concept for some of the state and national standards students are expected to learn? Can you think of other concepts needed to build a bridge between ideas in your science curriculum?

8. Describe how the card sort strategy can be used with this probe. How does the card sort strategy help the teacher visually collect data as students are discussing the probe? How does it promote learning as students share and explain their thinking?

9. Without going into biochemical details, how can you help elementary students understand that plants make sugars? What evidence can you use to help younger students understand that sugars are the food plants make?

10. When are students developmentally ready to connect the concept of food to two ideas: (1) atoms as building blocks for growth and repair of organisms and (2) the transfer of energy from the Sun? How does your curriculum or instructional materials address this developmental readiness?

PUTTING FORMATIVE ASSESSMENT INTO PRACTICE

1. What did you learn about your students' ideas from using the probe, "Is It Food for Plants?" Were you surprised by any of their answer choices and the rule or reasoning used to explain their thinking?

2. What evidence of partial understanding did you find in your students' responses? What evidence of common misconceptions related to food or photosynthesis did you find in your students' responses? What does this information reveal about the need for a bridging concept?

3. How is photosynthesis or needs of plants and animals typically taught in your science curriculum? What modifications will you make to your curriculum,

learning targets, instruction, or instructional materials based on your students' responses to this probe?

4. Did you provide an opportunity for your students to discuss and share their ideas with each other as they responded to the probe? Did you use an interactive strategy such as a card sort? If so, what did you observe about your students as they discussed their ideas with their peers?

5. What postassessment could you (or did you) give your students to determine the extent to which they now understand what plants use as food? What opportunity did you provide for your students to reflect on how their ideas may have changed?

6. Based on what you learned about your students' ideas related to the food plants make, what advice or suggestions do you have for your colleagues and future teachers?

GOING FURTHER

1. Read and discuss the Teacher Notes for the "Is It Food for Plants?" probe on pages 114–119 (Keeley, Eberle, and Tugel 2007). Pay particular attention to the Related Research and Suggestions for Instruction and Assessment sections.

2. An organism's need for food and the process of photosynthesis are addressed in *A Framework for K–12 Science Education* as core idea LS1.C: Organization for Matter and Energy Flow in Organisms pages 147–148 or online at *www.nap.edu/openbook.php?record_id=13165&page=147*. Read and discuss how this probe relates to what K–8 students should know about the food plants make and use.

3. The *Benchmarks for Science Literacy* explicitly address food for plants in the grades 6–8 essay and learning goals for 5E Flow of Matter and Energy *www.project2061.org/publications/bsl/online/index.php?chapter=5#E1*. Examine and discuss how *Benchmarks* addresses the scientific concept of food and how this information can be used to teach the bridging concept of food.

4. Examine the strand map "Flow of Energy Through Ecosystems" at *http://strandmaps.nsdl.org/?id=SMS-MAP-1422*. Discuss how the concept of food and ideas related to plant and animal needs for food progressively develop from grades K–2 to 6–8.

5. A framework to support the teaching of the topic "Flow of Matter and Energy in Ecosystems" is described in Koba, S. 2011. *Hard-to-teach science concepts: A framework to support learners, grades 3–5.* Arlington, VA: NSTA Press.

Chapter 17
Where Did the Water Go?

A *Framework for K–12 Science Education* gives teachers a new perspective on teaching and learning science by connecting three dimensions: Scientific and engineering practices, crosscutting concepts, and disciplinary core ideas (NRC 2012). The *Framework* committee points out that integrating these three dimensions into a system of standards, curriculum, and assessment is necessary in order to achieve the vision of the *Framework* and the *Next Generation Science Standards*. However, the committee also recognizes that "integrating the three dimensions in a coherent way is challenging and that examples of how it can be achieved are needed" (NRC 2012, p. 217). These examples should include ways to embed formative assessment into teaching and learning in order to monitor students' progress toward achieving interconnected goals for learning and to make the necessary adjustments to instruction that will promote deep, conceptual learning.

During formative assessment, standards-based teaching and learning are intertwined (Keeley 2008). Formative assessment not only informs teaching; it also supports learning across the three dimensions of *A Framework for K–12 Science Education*. These three dimensions are as follows:

Dimension 1: Scientific and Engineering Practices
These are the major practices scientists and engineers use to build an understanding of the world or design ways to solve problems. For example, asking questions, constructing explanations, and engaging in argument from evidence are a few of the important scientific practices for K–12 classrooms.

Dimension 2: Crosscutting Concepts
These are the concepts that have applicability across the disciplines of science. For example, patterns; structure and function; systems and system models; and energy and matter. Flows, cycles, and conservation are a few of the crosscutting concepts that help us understand both physical and living systems.

Dimension 3: Disciplinary Core Ideas
These describe the important content that K–12 students need to understand, use, and produce scientific knowledge. These core ideas are grouped in the disciplines of physical sciences; life sciences; Earth and space sciences; and engineering, technology, and applications of science.

Formative assessment is used to gather information about students' thinking related to core ideas as well as their ability to engage in scientific practices such as evidence-based argumentation. Let's examine how a teacher can use a formative assessment probe and formative assessment classroom techniques (FACTs) to uncover students' ideas, make instructional decisions, and foster conceptual learning. The assessment probe "Where Did the Water Come From?" (Figure 17.1) elicits students' preconceptions about condensation phenomena (Keeley, Eberle, and Dorsey 2008). The probe can be used with FACT claim cards to engage students in the scientific practice of engaging in argument from evidence. Claim cards are just that, cards on which students write a claim, in this case their probe answer. These assessment tools and techniques target the initial idea that water can be in the air around us in a form we cannot see called water vapor. The teacher can then use evidence of the students' understanding of this fundamental idea to further develop the crosscutting concept of the cycling of matter in the context of the water cycle.

Prior to teaching a sequence of lessons to help students develop a conceptual understanding of the processes of evaporation and condensation, the teacher gave students the probe and asked them to complete it individually. As she taught fifth-grade students, she did not expect them to use ideas about energy or the motion of molecules in their explanation. She was interested in how they explained the phenomenon macroscopically using their operational understanding of condensation. As she reviewed their answer choices and explanations, she noted that almost two-thirds of the class attributed the puddle to water from the inside of the container (choices B, D, and E). Some students attributed the phenomenon to coldness being changed into water, which was consistent with the research conducted by Bar and Travis (1991) described in the teacher notes for the probe. Some students correctly chose A and explained that evaporated water in the air touched the cold container and turned to a liquid. She took note of these students for the next day's activity.

The next day she had her students form small groups of six, making sure there was at least one student in each group who chose response A. She gave each group a set of six FACT claim cards (Keeley 2014). Students recorded their answer choices to the probe in claim format using a complete sentence (as instructed to do when they previously learned how to state a claim). For example, the card for response A stated: The puddle of water came from a gas in the air; the card for response B stated: The puddle of water came from melted ice inside the container, and so forth.

Each student then took turns reading the claim on their card to their group, and together the group decided whether they agreed or disagreed with each claim, supporting or

FIGURE 17.1.

The Where Did the Water Come From? Probe

Where Did the Water Come From?

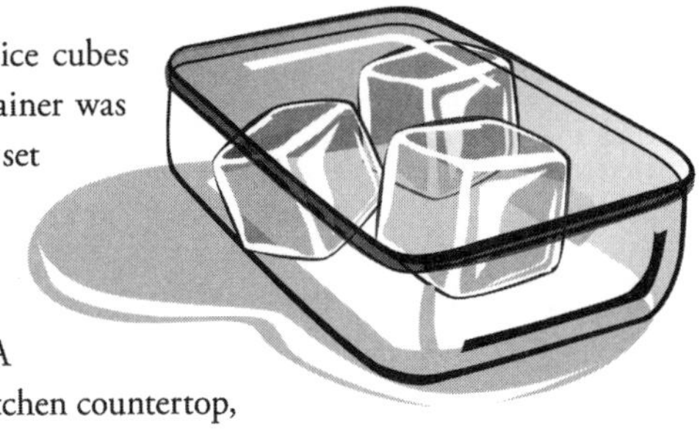

Latisha took a sealed, plastic container of ice cubes out of the freezer. The outside of the container was dry when she took it out of the freezer. She set the container on the counter. She did not open the container. Half an hour later she noticed the ice had melted inside the container. The container was full of water. A small puddle of water had formed on the kitchen countertop, around the outside of the container. Which best describes where the puddle of water came from?

A A gas in the air.

B Melted ice inside the container.

C Cold on the outside of the container.

D Condensation from water inside the container.

E Water that evaporated from inside the container.

F Cold changed hydrogen and oxygen atoms to water.

Describe your thinking about where the water came from. Provide an explanation for your answer.

refuting the claim with evidence. Some students pointed out why the evidence for some of the claims accepted by other members of the group was faulty, using reasoning from their everyday experiences and understanding of science concepts. After the small group discussion, each group was instructed to select the claim they could all agree on and construct an argument to present to the class as to why they thought it was the best claim. As the groups presented their claims and evidence to the whole class and engaged in argumentation with their peers, some students' ideas began to change or deepen. Eventually the whole class settled on the claim that the water came from a gas in the air and could explain the evidence that supported the claim. In essence, these students were engaging in the scientific practice of argumentation using claims and evidence (Dimension 1). At the same time they were socially constructing a core idea that would help them make sense of the phenomenon of water appearing on the outside of a cold object (Dimension 3).

This opportunity to develop conceptual understanding of a core idea through claims and evidence-based argumentation preceded the teacher's formal introduction of the terminology and processes of evaporation and condensation. She then used an additional probe, "Wet Jeans" (Keeley, Eberle, and Farrin 2005), and other FACTs to elicit further evidence of their conceptual understanding of evaporation and condensation.

The *Framework* notes that "students' understanding of crosscutting concepts should be reinforced by repeated use of them in the context of instruction in the disciplinary core ideas" (NRC 2012, p. 101). In turn, the crosscutting concepts (Dimension 2) can provide a connective structure to support students' understanding of disciplinary content (Dimension 3). Energy and Matter: Flows, Cycles, and Conservation is one of the crosscutting concepts. This crosscutting concept is described as "tracking fluxes of energy and matter into, out of, and within systems helps one understand the systems' possibilities and limitations" (NRC 2012, p. 84). As the teacher listened carefully to the students identify and explain the puddle and other phenomena, such as why wet dew forms on the grass in the morning, she then began to think about experiences that would help them use the crosscutting concept. She used information about her students' present understanding of evaporation and condensation to plan instruction that would help them understand the bigger idea about how water cycles between the atmosphere and the surface of the Earth.

Integrating across the three dimensions in the *Framework*—scientific practices, crosscutting concepts, and disciplinary core ideas—will present a new challenge to science teachers as they plan for curriculum, instruction, and assessment. As this example illustrates, the assessment probes in the *Uncovering Student Ideas in Science* series, used with formative assessment classroom techniques (FACTs), provide interesting and engaging contexts and strategies for linking these three dimensions while assessing and promoting deeper conceptual learning.

INTERNET RESOURCE

Uncovering Student Ideas in Science series
www.uncoveringstudentideas.org

REFERENCES

Bar, V., and A. Travis. 1991. Children's views concerning phase changes. *Journal of Research in Science Teaching* 28 (4): 363–382.

Keeley, P. 2008. *Science formative assessment: 75 practical strategies for linking assessment, instruction, and learning.* Thousand Oaks, CA: Corwin Press

Keeley, P. 2014. *Science formative assessment: 50 more practical strategies for linking assessment, instruction, and learning.* Thousand Oaks, CA: Corwin Press.

Keeley, P., F. Eberle, and C. Dorsey. 2008. *Uncovering student ideas in science, volume 3: Another 25 formative assessment probes, Volume 3.* Arlington, VA: NSTA Press.

Keeley, P., F. Eberle, and L. Farrin. 2005. *Uncovering student ideas in science, volume 1: 25 formative assessment probes, Volume 1.* Arlington, VA: NSTA Press.

National Research Council. 2012. *A framework for K–12 science education: Practices, crosscutting concepts, and core ideas.* Washington, DC: National Academies Press.

NSTA CONNECTION

Download the "Where Did the Water Come From?" probe at *www.nsta.org/SC1207*. Read the introduction to *Uncovering Student Ideas in Science, Volume 1: 25 Formative Assessment Probes* in the NSTA Science Store (*http://bit.ly/vaihtZ*).

• •

Where Did the Water Go?
Reflection and Study Guide

QUESTIONS TO THINK ABOUT AFTER YOU READ THIS CHAPTER

1. Summarize in your own words, the three dimensions of science learning: (1) disciplinary core ideas, (2) scientific and engineering practices, and (3) crosscutting concepts. In your teaching, do you place equal emphasis on all three or do you emphasize one more than the others?

2. Connecting the three dimensions of science learning—core disciplinary ideas, scientific and engineering practices, and crosscutting concepts—is a new way of teaching and assessing learning. Does this integration across dimensions present a challenge for your instruction and assessment?

3. What do you think it means to integrate the "three dimensions in a coherent way?"

4. Claim Cards are a formative assessment classroom technique (FACT). How would you describe this FACT? How can it be used with formative assessment probes?

5. Examine the best answer and the distracters (wrong answer choices) to this probe. What do you think underlies students' thinking for each of these answer choices?

6. Do you think students can use words like *evaporation* and *condensation* correctly and still lack conceptual understanding of the phenomenon? What does this tell us about the way we introduce vocabulary words into science?

7. How can the "Wet Jeans" probe be used to elicit further evidence of students' conceptual understanding of water cycle concepts?

8. How did the teacher connect the crosscutting concept of matter cycles (Energy and Matter) to the "Where Did the Water Come From?" probe?

9. What other condensation phenomena could you use to elicit students' ideas and provide an opportunity for them to engage in the practice of scientific argumentation?

10. *A Framework for K–12 Science Education* and *The Next Generation Science Standards* emphasize scientific and engineering practices as one of the three dimensions of learning science. One of these practices is engaging in argument from evidence. How does the teacher in this chapter support the use of this scientific practice in her classroom? (For more information about scientific and engineering practices visit NSTA's *NGSS* portal at *http://ngss.nsta.org*.)

PUTTING FORMATIVE ASSESSMENT INTO PRACTICE

1. What did you learn about your students' ideas from using the probe, "Where Did the Water Come From?" Were you surprised by any of their answer choices and the rule or reasoning used to explain their thinking?

2. Did you integrate the three dimensions of science learning as you used this probe? If so, how well do you think you accomplished the integration across dimensions?

3. Did you use the Claim Cards strategy? How well did it work with your students? Would you make any modifications to the strategy?

4. What instructional decisions did you make as you used the probe? Did students' responses indicate the need to address any gaps in your curriculum?

5. Based on your students' responses, how might you change the way science vocabulary is introduced and used when teaching and learning science?

6. Based on what you learned about your students' ideas related to condensation phenomena what advice or suggestions do you have for your colleagues and future teachers?

GOING FURTHER

1. Read and discuss the Teacher Notes for the "Where Did the Water Come From?" probe on pages 164–169 (Keeley, Eberle, and Dorsey 2008). Pay particular attention to the Related Research and Suggestions for Instruction and Assessment sections.

2. Examine the strand map "Weather and Climate" at *http://strandmaps.nsdl.org/?id=SMS-MAP-1422*. Discuss how water cycle ideas progressively develop from grades K–2 to 6–8. Notice how conceptual ideas are developed before terminology is introduced.

3. Learn more about the Claim Cards FACT and other techniques that support argumentation when using a formative assessment probe by reading and trying out the FACTs in the *Science Formative Assessment* series (Keeley 2008, 2014).

4. Read and discuss the teacher notes that accompany the mystery story "The Little Tent That Cried" in the *Everyday Science Mysteries* (Konicek-Moran 2008).

5. Read and discuss an article that describes how scientific modeling is used to represent and understand concepts of evaporation and condensation: Kenyon, L., C. Schwartz, and B. Hug. 2008. The benefits of scientific modeling. *Science and Children* 45 (2): 40–44.

Chapter 18

Confronting Common Folklore: Catching a Cold

Almost every child has experienced the sniffly, stuffy, and achy congestion of the common cold. In addition, many have encountered the "old wives tales" that forge a link between personal actions and coming down with this common respiratory infection. Much of this health folklore has been passed down from generation to generation (e.g., getting a chill or going outside in the winter with wet hair will cause you to get a cold) and affects the way children (and adults) view the cause of disease and its transmission. Even the common term used for this viral affliction, *a cold*, implies that weather or temperature has something to do with contracting this illness.

In their everyday conversations about health and sickness, children may form incorrect ideas about infectious disease transmission and often pass these common misconceptions on to their own children when they become adults. For example, it is not uncommon to hear an adult in the wintertime say to a child, "Don't go outside without your coat—you'll catch a cold." The formative assessment probe "Catching a Cold" (Figure 18.1, p. 131) is designed to elicit students' ideas about the common cold and find out whether they confuse the cause of this common infectious disease with prevalent myths about catching a cold or other factors that might contribute to weakening the human body's ability to ward off this familiar affliction (Keeley and Tugel 2009).

Using the Probe

Before using this probe, teachers should know that the common cold is an infection caused by a virus that is transmitted between two people, one who is contagious (infected with the virus) and one who picks up the virus. The *cause* is the virus (use the familiar word *germs* with younger children), and *transmission* is how it is spread (e.g., through contact with respiratory secretions). Other factors, such as getting a chill, may contribute indirectly to weakening the immune system, which fights off the virus. The important idea is that there is only one cause for the common cold: germs (specifically, a rhinovirus).

The distracters (wrong answer choices) on the probe are based on common myths, including the last one—the imbalance of body fluids (which should be removed for elementary students). This distracter was added for middle and high school students and dates back to historical beliefs about illnesses, which were treated by bleeding or purging. Because the incidence of contracting a cold rises during autumn when children return to school and are in close contact indoors with one another, this is a good time to use this probe. Elicit students' preconceptions about what causes a cold, and use their ideas as a springboard to differentiate between the cause of an infectious disease, the transmission of the disease, factors that contribute to good health, and unhealthy behaviors that can contribute to catching a cold (such as not washing your hands before eating).

Cold Conceptions

In the elementary grades, children develop an understanding of various science-related personal and societal health issues. While content related to cells, viruses, bacteria, and the immune system is not addressed until middle school, elementary children learn that there are very tiny things we cannot see called *germs*, and these germs can be spread from person to person. They learn about healthy habits such as good nutrition, exercise, keeping warm and dry, and getting enough sleep, but have difficulty distinguishing between factors that can affect good health and the causes of infectious diseases such as colds and flu. They are taught and told repeatedly to wash their hands in order to remove germs so that they are not passed on to others or taken into their body when their hands come in contact with their face or the food they eat. They learn to cover their coughs and sneezes in order to control transmission.

Even with this emphasis on controlling how germs can be passed from person to person and enter our body, children still fall back on common myths and misconceptions about what actually causes a cold. Some of the more common incorrect responses to this probe include *being wet, being wet and cold,* and *cold weather.* A look at the available research on children's conceptions of infectious disease helps explain why children also select these other factors as *causes,* even when they learn that germs are the only cause of colds and flu. Driver and colleagues (1994) points out that the word *cold* reinforces the connection to environmental causes, which helps explain why many children choose "cold weather." Many children lack a concept of microbes as agents of disease and attribute sickness to lifestyle behaviors. Even children who were interviewed and able to explain that diseases are caused by germs also believed people could catch a cold by getting cold and wet.

Children are not alone in holding these common misconceptions about colds. The medical folklore attached to the common cold is tenacious, even among adults. One of the benefits of using the *Uncovering Student Ideas in Science* probes is that teachers often uncover their own long-held misconceptions. Perhaps this probe has challenged some of your beliefs that were passed on to you in childhood. Now you no longer have to worry if you forget to bundle up when you go outside with your students at recess. As long as you don't come in direct contact with an infectious person's respiratory secretions, your likelihood of catching a cold won't increase!

FIGURE 18.1.

The Catching a Cold Probe

Catching a Cold

Have you ever been sick with a cold? People have different ideas about what causes a cold. Check off the things that cause you to "catch a cold."

___ having a fever

___ being wet

___ being wet and cold

___ germs

___ spoiled food

___ not getting enough sleep

___ lack of exercise

___ cold weather

___ dry air

___ imbalance of body fluids

Explain your thinking. Describe how people "catch a cold."

REFERENCES

Driver, R., A. Squires, P. Rushworth, and V. Wood-Robinson. 1994. *Making sense of secondary science: Research into children's ideas.* London and New York: RoutledgeFalmer.

Keeley, P., and J. Tugel. 2009. *Uncovering student ideas in science, volume 4: 25 new formative assessment probes.* Arlington, VA: NSTA Press.

NSTA CONNECTION

Download the "Catching a Cold?" probe at *www.nsta.org/SC1209*. Read the entire chapter and the introduction to *Uncovering Student Ideas in Science, Volume 1: 25 Formative Assessment Probes* in the NSTA Science Store (*http://bit.ly/vaihtZ*).

Confronting Common Folklore: Catching a Cold Reflection and Study Guide

QUESTIONS TO THINK ABOUT AFTER YOU READ THIS CHAPTER

1. What is meant by "common folklore" and how does it affect the ideas students bring to the science classroom?

2. What are some common misconceptions from folklore passed on from generation to generation you have encountered in your daily life?

3. Why do you think myths related to infectious diseases are so prevalent? Have you held on to any of these myths since childhood?

4. How does our every day language affect the way we think about scientific concepts? How does referring to a viral infection as a "cold" affect a child's thinking about the cause of an infectious disease?

5. How is the word *germs* used with younger children? What do you think the word *germs* means to them?

6. How might you modify this probe for the grade level you teach?

7. How could you use this probe in a health class or in health education?

8. Hand washing is a highly emphasized practice in schools. How can you connect this probe to helping students understand the importance of hand washing?

PUTTING FORMATIVE ASSESSMENT INTO PRACTICE

1. What did you learn about your students' ideas by examining their responses to the probe? Were folklore myths evident in your students' explanations? Were some answer choices more prevalent than others?

2. How did your use of this probe inform your curriculum or instruction?

3. Based on what you learned from using this probe with your students, what suggestions do you have for your colleagues and future teachers?

GOING FURTHER

1. Read and discuss the Teacher Notes for the "Catching a Cold" probe (Keeley and Tugel 2009, pp. 126–130). Pay particular attention to the Related Research and Suggestions for Instruction and Assessment sections.

2. Read and discuss the article: Pea, C., and D. Sterling. 2002. Cold facts about viruses. *Science Scope* 25 (3): 12–17.

3. Read and discuss the instructional considerations and learning goals for section 6E Physical Health in the *Benchmarks for Science Literacy* (AAAS 2009). *www.project2061.org/publications/bsl/online/index.php?chapter=6#E0*

Chapter 19
Talking About Shadows

A *Framework for K–12 Science Education* describes core ideas in Earth and space science that are central to the elementary science curriculum. By the end of grade 2, one of the ideas that all students should have an opportunity to develop is that patterns of the motion of the Sun, Moon, and stars can be observed, described, and predicted (NRC 2012). Shadow investigations provide an ideal context for children to collect and analyze data used to recognize and explain predictable patterns of the Sun's apparent path across the sky. Shadow investigations typically begin with a question such as, "How does our shadow change throughout the day?" Before beginning with the question that guides the students' investigation, consider starting with a formative assessment probe, such as "Me and My Shadow" (Figure 19.1, p. 136) to activate students' thinking and reveal claims that can be supported by or changed as a result of the evidence gathered through a shadow investigation (Keeley, Eberle, and Dorsey 2008). Furthermore, the unique format of this type of probe models scientific practices students should engage in when learning science content.

Friendly Talk

"Me and My Shadow" is a type of probe called a friendly talk probe (Keeley 2008). Many of the probes in the *Uncovering Student Ideas in Science* series are designed using this format to promote the importance of "science talk." A friendly talk probe is always set in the context of a group of people talking about a phenomenon. Each answer choice is given as a person stating a claim about the phenomenon. Students respond to the probe by choosing the person they most agree with and then explain why they agree.

The *friendly talk* probe format highlights an important form of communication in the elementary science classroom that parallels a central feature of science in the real world—the role of talk and argument. By modeling a scenario in which people engage in talking about their different ideas, the probe mirrors the importance of productive talk as a way for students to surface their initial claims and provide an explanation to support their initial claim. When used before developing an investigative question and launching into inquiry, the probe format provides an opportunity for students to practice oral skills of communication to articulate their thoughts and make their thinking visible to others. By agreeing or disagreeing with one of the characters in the probe, and providing an argument for why they agree or disagree, students are employing

> FIGURE 19.1.
> ## The Me and My Shadow Probe
> # Me and My Shadow
>
> Five friends were looking at their shadows early one morning. They wondered what their shadows would look like by the end of the day. This is what they said: 
>
> Jamal: "My shadow will keep getting longer throughout the day."
>
> Morrie: "My shadow will keep getting shorter throughout the day."
>
> Amy: "My shadow will keep getting longer until it reaches its longest point and then it will start getting shorter."
>
> Fabian: "My shadow will keep getting shorter until noon and then it will start getting longer."
>
> Penelope: "My shadow will stay about the same from morning to day's end."
>
> Which friend do you most agree with? _____
>
> Describe your thinking. Explain the reason for your answer.

communication skills used by scientists. As students choose the person they agree with in the probe and talk about their reasons for agreeing, the teacher uses "talk moves" to encourage productive classroom talk.

Table 19.1 shows six productive classroom talk moves from *Ready, Set, Science! Putting research to work in K–8 classrooms* (Michaels, Shouse, and Schweingruber 2008). Examples are included to show how teachers can help students clarify and expand their reasoning and arguments when using a *friendly talk* probe such as "Me and My Shadow."

Using the Probe

Teachers can use different talk formats with the *friendly talk probes*. Think-pair-share; partner talk; small-group discussion, and whole-class discussion are all ways to engage

TABLE 19.1.

"Talk Moves" and Examples

Talk Move (from *Ready, Set, Science!*)	Example of Talk Move from "Me and My Shadow"
Revoicing	"So let me see if I've got your thinking right. You're agreeing with Amy because _____?"
Asking students to restate someone else's reasoning	"Can you repeat in your own words what he just said about why he agrees with Jamal?"
Asking students to apply their own reasoning to someone else's reasoning	"Do you agree or disagree with her reason for agreeing with Morrie and why?"
Prompting students for further participation	"Would someone like to add on to the reasons why some of you chose Fabian?"
Asking students to explain their reasoning	"Why do you agree with Penelope?" or "What evidence helped you choose Fabian as the person you most agree with?" or "Say more about that."
Using wait time	"Take your time ... We'll wait."

students in talking and listening to each other's ideas and arguments. Productive classroom talk using a formative assessment probe before launching into an investigation also has the benefit of leading to deeper engagement in the content before and during the investigation. As students collect, analyze, and share their data, they compare their findings with their initial claims and evidence and become aware of the discrepancies between their own or others' ideas and the evidence gathered from the investigation. Students also learn that solar noon depends on the date and location.

The probe can be revisited again after students have had an opportunity to make sense of their data and use it to explain the apparent motion of the Sun across the daytime sky. This time the students have actual data from their investigation and a scientific explanation to support their reason for agreeing with Fabian. As they engage in talk and argument again with the same probe, the context of the probe, combined with the knowledge they constructed through their investigation, provides an opportunity for them to build stronger arguments.

Whether you use "Me and My Shadow" or any of the *friendly talk* probes in the *Uncovering Student Ideas in Science* series, consider using these types of formative assessment probes to not only gather information about your students' thinking, but to also encourage thinking through talk and argument. By seeing "others" in the probe talking about their ideas, students recognize the importance of sharing their thinking with their peers. This approach is not only integral to supporting content learning; it also helps students develop oral skills of communication and engage in methods of communication that are used by scientists in the real world.

REFERENCES

Keeley, P. 2008. *Science formative assessment: 75 practical strategies for linking assessment, instruction, and learning.* Thousand Oaks, CA: Corwin Press.

Keeley, P., F. Eberle, and C. Dorsey. 2008. *Uncovering student ideas in science, volume 3: Another 25 formative assessment probes.* Arlington, VA: NSTA Press.

Michaels, S., A. Shouse, and H. Schweingruber. 2008. *Ready, set, science! Putting research to work in K–8 classrooms.* Washington, DC: National Academies Press.

National Research Council. 2012. *A framework for K–12 science education: Practices, crosscutting concepts, and core ideas.* Washington, DC: National Academies Press.

NSTA CONNECTION

Download the "Me and My Shadow" probe at *www.nsta.org/SC1210*. Read the entire chapter and the introduction to *Uncovering Student Ideas in Science, Volume 1: 25 Formative Assessment Probes* in the NSTA Science Store (*http://bit.ly/vaihtZ*).

● ●

Talking About Shadows
Reflection and Study Guide

QUESTIONS TO THINK ABOUT AFTER YOU READ THIS CHAPTER

1. How do shadow investigations help students develop an understanding of patterns of motion in the Sun-Earth system?

2. Why is it a good idea to precede a scientific investigation with a formative assessment probe that elicits students' ideas?

3. How does the friendly talk format used in the "Me and My Shadow" probe mirror the way scientists discuss and communicate ideas?

4. How can you use the Talk Moves, described in Table 19.1, with the formative assessment probes? How does using talk moves facilitate more productive science talk in the classroom?

5. What are some of the different formats and structures you use to support science talk in your classroom?

6. *A Framework for K–12 Science Education* and *The Next Generation Science Standards* emphasize scientific and engineering practices as one of the three dimensions of learning science. One of these practices is engaging in argument from evidence. How can the "Me and My Shadow" probe be revisited after students have collected data to engage in this scientific practice? (For more information about scientific and engineering practices visit NSTA's *NGSS* portal at *http://ngss.nsta.org*.)

7. How can the "Me and My Shadow" probe be used to help students use the crosscutting concept of patterns? (For more information about crosscutting concepts visit NSTA's *NGSS* portal at *http://ngss.nsta.org*.)

PUTTING FORMATIVE ASSESSMENT INTO PRACTICE

1. What did you learn about your students' ideas about shadows? Were you surprised by any of their answer choices and explanations?

2. How did the formative assessment probe engage and motivate your students as they launched into investigating a scientific question?

3. What evidence did you see of students changing their initial ideas as they carried out their shadow investigation?

4. Which talk moves did you tend to use most during the science talk? Which talk moves do you need more practice in using?

5. To what extent do you think your students engaged in the scientific practice of argumentation? What aspects of argumentation do your students need more practice with during science talk?

6. To what extent did your students use the crosscutting concept of Patterns to examine, analyze, and explain their data?

7. Based on what you learned about your students' ideas related to shadows and how they change throughout the day, what advice or suggestions do you have for your colleagues and future teachers?

GOING FURTHER

1. Read and discuss the Teacher Notes for the "Me and my Shadow" probe on pages 186–190 (Keeley, Eberle, and Dorsey 2008). Pay particular attention to the Related Research and Suggestions for Instruction and Assessment sections.

2. There are several additional probes that target children's ideas about the Sun-Earth system in *Uncovering Student Ideas about Astronomy* (Keeley and Sneider 2012). The teacher notes for these probes provide additional background infor-

mation on common misconceptions and instructional suggestions for making formative decisions.

3. Patterns of motion in the Sun-Earth system is included in the *NGSS* and *A Framework for K–12 Science Education* under Core Idea ESS1-Earth's Place in the Universe. Read pages 173–176 or online at *www.nap.edu/openbook.php?record_id=13165&page=145* and discuss how this probe relates to how students can predict and explain the changes in shadow length throughout the day.

4. Read and discuss the article: Barrows, L. 2007. Bringing light onto shadows. *Science and Children* 44 (9): 43–45.

5. Read and discuss the Teacher Notes to the everyday science mystery story "Where Are the Acorns?" in the *Everyday Science Mysteries* (Konicek-Moran 2008).

6. Watch and discuss the NSTA-archived web seminar on the *NGSS* core idea Earth's Place in the Universe at: *http://learningcenter.nsta.org/products/symposia_seminars/NGSS/webseminar31.aspx.*

7. Watch and discuss the NSTA-archived web seminar on the *NGSS* crosscutting concept of Patterns at: *http://learningcenter.nsta.org/products/symposia_seminars/NGSS/webseminar19.aspx.*

Chapter 20

Birthday Candles: Visually Representing Ideas

Several studies have been conducted on children's pervasive misconceptions related to light transmission and vision. Over half a century ago, Piaget's research showed that some children tend to think an object is seen when something passes from the eye to an object, rather than the reverse (Piaget 1974). Several later studies used interviews and children's drawings to analyze their ideas about the connection between light sources, objects, and vision (Fetherstonhaugh and Treagust 1992; Osborne et al. 1990). These studies revealed commonly held ideas, such as it is the person's act of looking at an object that allows one to see it, light is necessary to illuminate an object but the light remains around the objects rather than traveling to the eye, and that light stays close to a light source. Interestingly, one analysis of younger children's drawings showed that they were far more similar to one another than the ideas of older children, whose drawings were more varied (Osborne et al. 1990).

A Framework for K–12 Science Education's Core Idea PS4: Waves and Their Applications in Technologies for Information Transfer and Component Idea PS4.B: Electromagnetic Radiation include concepts related to light transmission that help elementary students develop an understanding of this familiar, yet mysterious phenomenon (NRC 2012). Some of the ideas related to light that children should develop by the end of grade 2 include: light travels from place to place, objects can be seen only when there is light available to illuminate them, and some hot objects give off their own light. By the end of grade 5 children build upon these ideas by learning that light can travel through space and that light must travel from an object to the eye in order to see the object.

The notion that something can be seen when light from the object or light source enters our eye is addressed at the grades 3–5 level in the *Framework* (Component Idea PS4.B). In most states, currently enacted elementary science learning goals address the connection between light and vision at the middle school level. This is an example of one of the content placement transitions from the middle grades to the elementary level that schools and curriculum developers will face in implementing the *Next Generation Science Standards.* Introducing this idea earlier will help students continually build on and revise their knowledge about light as they move from one grade span to the next. The formative assessment probe "Birthday Candles" (Figure 20.1, p. 143) was developed to elicit students' ideas about how light travels from its source (Keeley, Eberle, and

Farrin 2005). With further probing, "Birthday Candles" can also reveal students' ideas about the connection between light and vision.

Using the Probe

An important feature of formative assessment, sometimes overlooked when using the probes in the *Uncovering Student Ideas in Science* series, is the use of visual representations to support students' explanations. Researchers use children's drawings to gain insight into children's thinking. Likewise, teachers can also use their students' drawings to examine their students' ideas. Elementary students' drawings, in combination with their written or verbal explanations, provide an additional window into their thinking that can be used by the teacher to formatively plan instruction while simultaneously promoting learning. The very act of making a drawing requires a child to surface, examine, and clarify their ideas in order to communicate their thinking visually.

For example, a fifth-grade teacher using the "Birthday Candles" probe extended the explanation part of the probe by asking students to include a drawing that shows how someone can see the light from the candles. Figure 20.2 (p. 144) is an example of a fifth-grade student's explanation for the "Birthday Candles" probe. Figure 20.3 (p. 144) shows the representation the student used to support his explanation. The drawing reinforces this student's notion of the "halo effect" surrounding the light source with the distance traveled by the light from the candles dependent upon how strong the light is. The small lines show light moving outward from the flames.

Teachers can ask students to individually draw their own picture to represent their thinking or they may use small-group whiteboarding. Whiteboarding is a formative assessment classroom technique (FACT) that involves a small group of 3–4 students in using a desktop whiteboard and dry erase markers to collaboratively represent the group's best thinking (Keeley 2008). The advantage in using whiteboards is that a drawing can be changed as students accept, discard, or modify their ideas while sharing their thinking and listening to the alternative ideas of others in their group.

Whether students create their own individual representation to support their explanation, such as the example in Figure 20.3, or small groups create a joint whiteboard representation, sharing their visually supported explanations with the whole class further promotes learning through formative assessment by providing students with an opportunity to practice communication skills and give and receive feedback to their peers. During the presentations, the teacher gathers formative data from the presentations and feedback to further plan instruction that will move students toward three key ideas in the elementary grade band endpoints of the *Framework* under the core idea, PS4.B: Electromagnetic

FIGURE 20.1.
The Birthday Candles Probe
Birthday Candles

Imagine you are at a birthday party. A birthday cake with candles is put on a table in the middle of a room. The room is very large. You are standing at the end of the room, 10 meters away from the cake. You can see the candles. Circle the reponse that best describes how far the light from the candles traveled in order for you to see the flames.

A The light stays on the candle flames.

B The light travels a few centimeters from the candle flames.

C The light travels about 1 meter.

D The light travels about halfway to where you are standing.

E The light travels all the way to where you are standing.

Describe your thinking. Provide an explanation for your answer.

Radiation: (1) hot objects give off light (e.g., flame from a burning candle), (2) light travels outward from a source to our eyes unless it is blocked by an object, and (3) we see objects or light sources because the light reflected from the object or given off by the light source travels to our eyes. The teacher may note evidence of alternative ideas that seem to be common among the students' presentations or feedback that will need to be addressed in subsequent lessons as well as indications of student understanding.

During the presentations the teacher also has an opportunity to provide formative feedback on students' visually supported explanations and probe further to elicit the key content ideas. For example, the teacher might ask the student who supported his "halo" explanation with the drawing in Figure 20.3 to add to his drawing a person standing at the end of the room. The teacher would then ask the student to use the drawing to show how that person would see the light around the candles, as drawn in his picture. The teacher would be carefully listening to the students' alternative ideas; or in some cases,

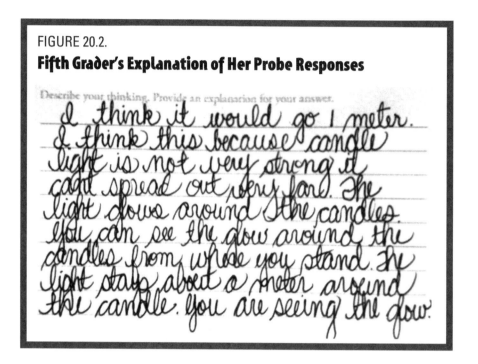

FIGURE 20.2.

Fifth Grader's Explanation of Her Probe Responses

Describe your thinking. Provide an explanation for your answer.

I think it would go 1 meter. I think this because candle light is not very strong it cant spread out very far. The light glows around the candles. You can see the glow around the candles from where you stand. The light stays about a meter around the candle. You are seeing the glow.

note how the student changes his explanation as he further thinks visually.

The adage, "A picture is worth a thousand words," certainly holds true for formative assessment. As you use the *Uncovering Student Ideas in Science* probes in your classroom, think about opportunities for students to visually represent their thinking. Not only will you gather additional data about your students' understanding of science concepts, "going visual" will help your students strengthen their explanations and enhance learning in the process.

FIGURE 20.3.

Fifth Grader's Drawing of His Thinking About Light

REFERENCES

Fetherstonhaugh, T., and D. Treagust. 1992. Students' understanding of light and its properties: Teaching to engender conceptual change. *Science Education* 76 (6): 653–672.

Keeley, P. 2008. *Science formative assessment: 75 practical strategies for linking assessment, instruction, and learning.* Thousand Oaks, CA: Corwin Press.

Keeley, P., F. Eberle, and L. Farrin. 2005. *Uncovering student ideas in science, volume 1: 25 formative assessment probes.* Arlington, VA: NSTA Press.

National Research Council (NRC). 2012. *A framework for K–12 science education: Practices, crosscutting concepts, and core ideas.* Washington, DC: National Academies Press.

Osborne, J., P. Black, S. Smith, and J. Meadows. 1990. *Light research report.* Primary SPACE Project. Liverpool, England: Liverpool University Press.

Piaget, J. 1974. *Understanding causality.* New York: W.W. Norton.

NSTA CONNECTION

Download the "Birthday Candles" probe at *www.nsta.org/SC12010*. Read the entire chapter and the introduction to *Uncovering Student Ideas in Science, Volume 1: 25 Formative Assessment Probes* in the NSTA Science Store (*http://bit.ly/vaihtZ*).

● ●

Birthday Candles:Visually Representing Ideas Reflection and Study Guide

QUESTIONS TO THINK ABOUT AFTER YOU READ THIS CHAPTER

1. What are some of the misconceptions children have about transmission of light and how one sees an object?

2. What are some of the ideas addressed at the K–6 level that students are expected to learn related to the concept of light transmission?

3. How is an explanation of how light travels related to an explanation of how we see objects? How does the probe "Birthday Candles" address both of these concepts? What other context, besides candles on a birthday cake, could you use to further probe students' ideas?

4. How can combining children's drawings with their written explanations provide greater insight into children's thinking? How does the representation in Figure 20.3 enhance the explanation in Figure 20.2?

5. What are some ways you can use children's drawings to learn more about their ideas in science?

6. Why is it important for students to share their drawings with their peers? How is feedback used with visual representations?

7. How does the adage, "A picture tells a thousand words" relate to formative assessment as assessment for learning?

8. *A Framework for K–12 Science Education* and *The Next Generation Science Standards* emphasize scientific and engineering practices as one of the three dimensions of learning science. One of these practices is Obtaining, Evaluating, and Communicating information. How does combining visual representations with the written or oral explanations support this practice? (For more information about scientific and engineering practices visit NSTA's *NGSS* portal at: *http://ngss.nsta.org*.)

PUTTING FORMATIVE ASSESSMENT INTO PRACTICE

1. What did you learn about your students' ideas after using this probe? Were you surprised by any of their answer choices and explanations?

2. Did your students include a drawing with their explanation? How did their drawings provide additional evidence of students' ideas?

3. How would you categorize your students' ideas related to this probe? Were their ideas similar to the research-identified commonly held ideas described in the chapter? Which one(s) were most prevalent among your students?

4. Did you find evidence of partial understanding? How could you build off these partial understandings to fully develop the scientific idea?

5. Where do you think your students' ideas about how light travels originated? Were you able to uncover any evidence that revealed the source of their misunderstanding? How could you use this knowledge to inform your teaching?

6. After analyzing your students' responses, what curricular or instructional decisions did you make?

7. How did students receive and give feedback on their ideas related to this probe? How did feedback play an integral role in their learning?

8. Based on what you learned about your students' ideas related to how light travels, what advice or suggestions do you have for your colleagues and future teachers?

GOING FURTHER

1. Read and discuss the Teacher Notes for the "Birthday Candles" probe on pages 38–41 (Keeley, Eberle, and Farrin 2005). Pay particular attention to the Related Research and Suggestions for Instruction and Assessment sections.

2. How light travels and the role of light in how we see objects is included in the *NGSS* and *A Framework for K–12 Science Education* under Core Idea PS4.B: Electromagnetic Radiation. Read pages 133–136 or online at *www.nap.edu/openbook. php?record_id=13165&page=133* and discuss how this probe relates to the ideas students are expected to understand at their grade level.

3. Examine the probes and teacher notes for "Can It Reflect Light?" and "Apple in the Dark" for further examples of assessments that connect the role of light in how we see objects (Keeley, Eberle, and Farrin 2005).

4. Watch and discuss the NSTA-archived web seminar on the *NGSS* scientific and engineering practice of obtaining, evaluating, and communicating information at *http://learningcenter.nsta.org/products/symposia_seminars/NGSS/webseminar12.aspx*.

5. Read and discuss an article that describes how modeling through use of representations is used to think through and communicate ideas: Kenyon, L., C. Schwartz, and B. Hug. 2008. The benefits of scientific modeling. *Science and Children 45* (2): 40–44.

6. The Private Universe Project in Science has a series of online streaming videos that address students' ideas in science. Watch and discuss video workshop #5 "Can We Believe Our Own Eyes" to learn more about students' commonly held misconceptions about light and the role of light in how we see objects. This series can be accessed *at www.learner.org/resources/series29.html* (Harvard-Smithsonian Center for Astrophysics 1995).

Chapter 21

Mountain Age: Creating a Classroom Profile

Earth science concepts, such as the weathering of landforms, can be particularly challenging for elementary students. One of the reasons for this difficulty is the inability of some young children to comprehend long periods of geologic time in which changes occur to landforms. Another difficulty lies in the literal interpretations students make of visual information. The formative assessment probe "Mountain Age" (Figure 21.1) can be used to elicit students' ideas about processes that affect the shape

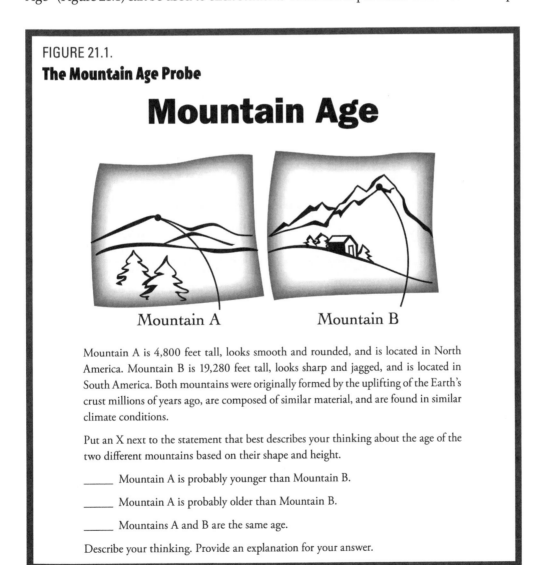

FIGURE 21.1.

The Mountain Age Probe

Mountain Age

Mountain A   Mountain B

Mountain A is 4,800 feet tall, looks smooth and rounded, and is located in North America. Mountain B is 19,280 feet tall, looks sharp and jagged, and is located in South America. Both mountains were originally formed by the uplifting of the Earth's crust millions of years ago, are composed of similar material, and are found in similar climate conditions.

Put an X next to the statement that best describes your thinking about the age of the two different mountains based on their shape and height.

_____ Mountain A is probably younger than Mountain B.

_____ Mountain A is probably older than Mountain B.

_____ Mountains A and B are the same age.

Describe your thinking. Provide an explanation for your answer.

of mountains (Keeley, Eberle, and Farrin 2005). There is no "right answer" to this probe because determining the relative age of mountains involves multiple interacting factors and cannot be determined solely by visual cues. However, the intent of this probe is to reveal whether students recognize the role of weathering in shaping the features of mountains.

A Framework for K–12 Science Education identifies and describes the core ideas that were used to inform the development of the *Next Generation Science Standards* (NRC 2012). Core Idea ESS2, Earth Systems, centers around the question "How and why is Earth constantly changing?" By the end of grade 5, students should know that "water, ice, wind, living organisms, and gravity break rocks, soils, and sediments into smaller particles and move them around" (NRC 2012, p. 181). Students can see the effects of these weathering and erosion agents, but observing the actual process is more difficult due to the scale of landforms and the long periods of time over which these processes occur. As a result, students try to connect Earth processes and features of landforms to other familiar processes or intuitive rules for making sense of phenomena, such as smaller–younger, taller–older.

A key part of formative assessment in science is uncovering the ideas students use to make sense of phenomena such as the changing shape of mountains. Creating a classroom profile for a formative assessment probe is a way for teachers to record, analyze, and share data with colleagues for the purpose of planning instruction that targets students' commonly held ideas about phenomena. It also provides a written record of students' misconceptions or partially formed ideas that the teacher can refer to in monitoring for conceptual change over time.

The first step in creating a classroom formative assessment profile of the probe "Mountain Age" is to scan through the student responses and sort by the three answer choices. Next, take each answer choice and look for commonalities in the students' explanations. Examine the similarities in reasoning and come up with a descriptor that can be used to group the similar ideas for that answer choice. For example, 10 students wrote explanations similar to the one shown in Figure 21.2. They may have worded their explanations differently, but all fit the descriptor "smaller mountains are younger." Two responses were similar to this, but they also added the idea that older mountains were more jagged, as seen in the Figure 21.3 example. Because these responses included additional information about the students' reasoning, they are grouped under the descriptor, "Smaller mountains are younger and not as jagged." Sometimes students merely parrot back the answer choice. This type of response, which does not give any insight into the students' reasoning, is recorded as "Repeats answer choice with no explanation."

The number of students with explanations that fit a descriptor is recorded in the third column and totaled for the overall answer choice, as seen in Table 21.1 (p. 152).

FIGURE 21.2.
"Smaller Mountains Are Younger" Descriptor

Describe your thinking. Provide an explanation for your answer.

I think mountain A is Probably younger than mountain B Because mountain A looks smaller than moutainB

FIGURE 21.3.
"Smaller Mountains Are Younger and Not as Jagged" Descriptor

Describe your thinking. Provide an explanation for your answer.

A is smaller then B. B is bigger Than A. B is more mature and better Then A, So HA. A also hasn't gotten any Jagged edges yet.

Teachers can also opt to include students' names or initials in order to track individual thinking. At a glance the teacher can see the similarities and differences in students' thinking. The profile also provides data that will help the teacher probe further or differentiate instruction for a group of students with similar misconceptions. For example, the teacher might want to probe further to find out if the two students with idea #6 have a beginning concept of weathering and erosion, or think of a way to help the four students with idea #7 recognize that mountain building happens throughout Earth's geologic history.

This exercise in grouping similar responses to create classroom formative assessment profiles is not only helpful for organizing formative assessment information on students' thinking, it is also a form of embedded professional development situated in real classroom data. Teachers can share and discuss the formative assessment profiles in their professional learning communities, thereby building the capacity of a collaborative group to examine student thinking and discuss instructional strategies for addressing students' ideas. Preservice and beginning teachers can develop and use the profiles to

TABLE 21.1.

Classroom Profile for the Mountain Age Probe, Grade 4

Probe Answer Choices	Students' Ideas	Number of Students
Mountain A is probably younger than Mountain B.	"Smaller mountains are younger."	10 (AB, KL, HV, NT, AS, PS, ATS, PT, JF, LC)
	"Smaller mountains are younger and not as jagged."	2 (TR, DED, KN)
	"Mountain A is still growing."	3 (FA, FL, JD, KK)
	Repeats answer choice with no explanation.	1 (PR)
		Total: 16
Mountain A is probably older than Mountain B.	"It's spread out more so its older"	1 (KJ)
	"Older mountains are more worn down."	2 (CC, ZH)
		Total: 3
Mountains A and B are the same age.	"All mountains on Earth formed at the same time"	4 (ML, GN, MM, BI)
	"Can't tell from a picture."	3 (OL, SRS, TT)
		Total: 7

deepen their understanding of children's preconceptions and see the value of formative assessment for making instructional decisions and monitoring for conceptual change. The profiles (with students' names removed) can also be shared with students. In a formative assessment–centered classroom, students show great interest in knowing what other students think and how their classmates' thinking compares to their own. Overall, classroom formative assessment profiles provide an easily visualized picture of classroom thinking about core ideas in science that support data-driven instructional decisions.

REFERENCES

Keeley, P., F. Eberle, and L. Farrin. 2005. *Uncovering student ideas in science, volume 1: 25 formative assessment probes.* Arlington, VA: NSTA Press.

National Research Council (NRC). 2012. *A framework for K–12 science education: Practices, crosscutting concepts, and core ideas.* Washington, DC: National Academies Press.

INTERNET RESOURCES

Uncovering Student Ideas in Science series
    www.uncoveringstudentideas.org

NSTA CONNECTION

Download the "Mountain Age" probe at *www.nsta.org/SC 1212*. Read the entire chapter and the introduction to *Uncovering Student Ideas in Science, Volume 1: 25 Formative Assessment Probes* in the NSTA Science Store (*http://bit.ly/vaihtZ*).

● ●

Mountain Age: Creating a Classroom Profile Reflection and Study Guide

QUESTIONS TO THINK ABOUT AFTER YOU READ THIS CHAPTER

1. Why are some Earth science concepts so difficult for young children to conceptualize?

2. Why is there no one right answer to the "Mountain Age" probe? How is uncovering student ideas more important than getting the "right answer" when using a probe such as "Mountain Age"?

3. How does this probe address the essential question, "How and why is Earth constantly changing?" What ideas does this probe elicit that address this question?

4. How do students connect their prior ideas and experiences to this probe? What intuitive rule do students use to think about the age of the two mountains in the probe?

5. What is a classroom profile and how is it used in formative assessment?

6. How can creating a classroom profile be used for professional learning?

7. Examine the fourth-grade classroom profile in Table 21.1 for the "Mountain Age" probe. If these were your students, what would this profile tell you? How would this data inform your next steps?

PUTTING FORMATIVE ASSESSMENT INTO PRACTICE

1. What did you learn about your students' ideas after using this probe? Were you surprised by any of their answer choices and explanations?

153

2. How would you categorize your students' ideas related to this probe? Did you find evidence of students using the intuitive rule, more A–more B, as described in Chapter 2?

3. Did you create a classroom profile of your data? Describe the process of creating your classroom profile. How would you share and discuss your classroom profile with your colleagues?

4. Did your classroom profile reveal how certain groups of students responded (for example, high, middle, and low achieving)? English language learners?

5. How could you work together with your colleagues to design instruction that will address students' ideas related to this probe?

6. Based on what you learned about your students' ideas related to weathering of mountains, what advice or suggestions do you have for your colleagues and future teachers?

7. Based on what you learned about the value of creating a classroom profile, what advice or suggestions do you have for your colleagues and future teachers?

GOING FURTHER

1. Read and discuss the Teacher Notes for the "Mountain Age" probe on pages 170–175 (Keeley, Eberle, and Farrin 2005). Pay particular attention to the Related Research and Suggestions for Instruction and Assessment sections.

2. Weathering of landforms is included in the *NGSS* and *A Framework for K–12 Science Education* under Core Idea ESS1.C: The History of Planet Earth. Read pages 177–179 or online at *www.nap.edu/openbook.php?record_id=13165&page=177* and discuss how this probe relates to the ideas students are expected to understand at their grade level.

3. The website Tools for Ambitious Science Teaching at *http://tools4teachingscience. org* has some excellent resources and protocols for examining student thinking.

4. CIEST (Collaborative Inquiry into Examining Student Thinking) is a protocol that can be used with formative assessment probes to examine student thinking. This protocol and materials for facilitating its use with probes is described in Mundry, S., P. Keeley, and C. Landel. 2010. *A leader's guide to science curriculum topic study*. Thousand Oaks, CA: Corwin Press.

5. Research how classroom profiles can be used within professional development strategies such as lesson study, collaborative inquiry into examining student thinking, critical friends groups, study groups, professional learning communities, and action research.

Chapter 22
Using the P-E-O Technique

O bserving how objects sink or float in water is an elementary precursor to developing explanations in later grades that involve an understanding of density and buoyancy. Beginning as early as preschool, elementary students engage in activities that encourage them to predict whether an object will sink or float when placed in water, and then test their predictions by making observations. Core Idea PS1.A from *A Framework for K–12 Science Education* (NRC 2012) states that by the end of grade 2 students should know that matter can be described by its observable properties and that different properties are suited to different purposes. Children often base their predictions about whether an object or material will float or sink on the observable properties of the object they are testing such as its size, felt weight, heaviness for its size, or shape. Therefore it is important for elementary students to have multiple experiences describing the properties of objects and materials, predicting whether they will float or sink, supporting their predictions with explanations that use the properties of the objects or materials as evidence, and testing their predictions.

The "Solids and Holes" probe (Figure 22.1, p. 157) is designed to find out whether students think a solid object will float differently if holes are poked all the way through it (Keeley, Eberle, and Tugel 2007). In a study by Grimillini, Gandolfi, and Pecori Balandi (1990) of children's ideas related to buoyancy, they found that children take into account four factors when considering how objects float: (1) the role played by material and weight; (2) the role played by shape, cavities, and holes; (3) the role played by air; and (4) the role played by water. All of these factors have been revealed in students' responses to this probe when using the P-E-O formative assessment classroom technique (FACT) (Keeley 2008).

To begin the P-E-O probe, the teacher provides students with a solid block of material that floats in water (e.g., wood, Styrofoam, or a piece of orange peel). Each block of material is the same size and shape. Students place their material in a cup of water and observe how it floats. P-E-O begins with a prediction (the "P" part of P-E-O) that draws upon students' prior knowledge and experiences. Students predict what they think would happen if holes were poked all the way through the material as shown on the illustration in the probe. Students select their prediction from each of the answer choices provided on the probe and write an explanation (the "E" part of P-E-O) to support their reasoning.

In small groups or as a whole-class discussion, students share their predictions and engage in the scientific practice of argumentation to defend their reasoning. The teacher listens carefully for evidence of understanding or misconceptions related to how objects float. For example, most students select (A) it will sink, based on their knowledge of what happens when a boat gets a hole in it. Other students will choose (B) it will barely float, with the reasoning that the water fills in the holes and weighs it down. Some students choose (D) it will neither sink nor float (it "flinks"!) due to their belief that the water will fill the holes and start to sink the object and then air will get into the holes and it will rise upward, repeating the cycle. After students have had an opportunity to share their thinking, they may reconsider their initial ideas and change their prediction based on the plausible arguments of others. The next step is to launch into an investigation to test their initial or revised predictions by making observations. This is the "O" part of the P-E-O probe.

Students are provided with tools and safety instructions for drilling or poking holes in their material (or the teacher can provide them with a new block with holes already made for them). Students place their hole-ridden object in the water and observe. Be prepared for students who will not accept what they observe. The idea that an object that has holes in it will sink is so strongly held that some students are convinced that there are not enough holes in the object or the holes are not big enough! Instead of correcting them or ending the investigation and giving students the answer, let the students make more or bigger holes. Eventually they will see that the material still floats the same way as it did before the holes were poked through it. This leads to the most important part of the P-E-O probe—revising an explanation when observations do not match the prediction. Now students work together in small groups and as a class to come up with an explanation that supports the observation. Eventually some students will recognize that all they did was take out some of the material and change its shape. It was still the same floating material. It is at this point that the teacher can build on the students' revised explanation to explain why the hole-ridden material floats the same way as the solid material and does not sink like a boat, partially sink, or "flink" (as described by the distracter answer choices). The air in the hull of a boat or in a hollow object is replaced by water when a hole is punched in it, thus making it sink. Boats and hollow objects include air and a solid material. The material students tested did not contain air and only involved one material. (Note: Later, when students learn about density in middle school, they will learn that the mass-volume relationship stays the same when holes are punched through a solid material).

What makes the P-E-O strategy an important technique for formative assessment is that it provides teachers with insights into the misconceptions students hold prior to engaging

Figure 22.1.
The Solids and Holes Probe

Solids and Holes

Lance had a thin, solid piece of material. He placed the material in water and it floated. He took the material out and punched holes all the way through it. What do you think Lance will observe when he puts the material with holes back in the water? Circle your prediction.

A It will sink.

B It will barely float.

C It will float the same as it did before the holes were punched in it.

D It will neither sink nor float. It will bob up and down in the water.

Explain your thinking. Describe the "rule" or reasoning you used to make your prediction.

in inquiry that are then used to facilitate the learning process. It also promotes learning through a process of conceptual change that involves important scientific practices contained in the *Next Generation Science Standards*, such as planning and carrying out investigations, analyzing and interpreting data, constructing evidence-based explanations, and engaging in scientific argumentation to defend one's ideas. As you look through the probes in the *Uncovering Student Ideas in Science* series, identify ones that can be used with the P-E-O technique to provide a constructivist approach to using the scientific practices for support of students' conceptual understanding.

REFERENCES

Grimillini, T., E. Gandolfi, and B. Pecori Balandi. 1990. Teaching strategies and conceptual change: Sinking and floating at elementary school level. Paper presented at the Australian Science Education Research Association Conference, Melbourne, Australia.

Keeley, P. 2008. *Science formative assessment: 75 practical strategies for linking assessment, instruction, and learning.* Thousand Oaks, CA: Corwin Press.

Keeley, P., F. Eberle, and J. Tugel. 2007. *Uncovering student ideas in science, volume 2: 25 more formative assessment probes.* Arlington, VA: NSTA Press.

National Research Council (NRC). 2012. *A framework for K–12 science education: Practices, crosscutting concepts, and core ideas.* Washington, DC: National Academies Press.

INTERNET RESOURCES

Uncovering Student Ideas in Science series
   www.uncoveringstudentideas.org

NSTA CONNECTION

Download the "Solids and Holes" probe at *www.nsta.org/SC1301*. Read the entire chapter and the introduction to *Uncovering Student Ideas in Science, Volume 1: 25 Formative Assessment Probes* in the NSTA Science Store (*http://bit.ly/vaihtZ*).

• •
Using the P-E-O Technique
Reflection and Study Guide

QUESTIONS TO THINK ABOUT AFTER YOU READ THIS CHAPTER

1. What are some of the research-identified factors children take into account to think about whether an object floats or sinks? How is it helpful to the teacher to be aware of these factors before planning floating and sinking activities?

2. What does P-E-O stand for and how is it used as a formative assessment classroom technique (FACT)?

3. P-O-E is a similar technique in which students make a prediction about a discrepant event, demonstration, or phenomenon, observe what happens, and then construct an explanation. How does P-O-E differ from P-E-O? What is the advantage of using P-E-O for formative purposes?

4. Examine the answer choices to the "Solids and Holes" probe. Explain what you think each answer choice is designed to reveal about students' thinking and the experiences or prior knowledge that may affect their answer choice.

5. Students often confuse the context of this probe with mixed density problems in which one of the objects is filled with air. What is different about this context? What do students need to recognize in order to differentiate this context from one such as a hollow ball floating in water?

6. P-E-O often ends up as P-E-O-E. When do students add the second "E"?

7. What is the advantage of using the P-E-O technique with students? Why not merely correct them if they choose the "wrong" answer?

8. What other probes in the *Uncovering Student Ideas in Science* series could you use with the P-E-O technique?

9. *A Framework for K–12 Science Education* and *The Next Generation Science Standards* emphasize a set of eight scientific and engineering practices as one of the three dimensions of learning science. Which scientific practices are students using with the P-E-O technique? (For more information about scientific and engineering practices visit NSTA's *NGSS* portal at *http://ngss.nsta.org*.)

PUTTING FORMATIVE ASSESSMENT INTO PRACTICE

1. What did you learn about your students' ideas after using this probe? Were you surprised by any of their answer choices and explanations?

2. How would you categorize your students' ideas related to this probe? Did you find evidence of students considering any of the four factors that affect floating that were described in the chapter?

3. What instructional decisions did you make after analyzing your students' responses?

4. Sometimes students hold on to their ideas so tenaciously that they may not accept what they observe. Sometimes students will say the holes need to be bigger or there needs to be more holes. These are examples of ways students accommodate their strongly held beliefs that conflict with what they actually observed. Did you notice any of your students holding on to their initial ideas, even when the evidence from their observations was presented? How did your students finally reconcile their observation?

5. What kinds of information did you gather from using this probe and P-E-O technique that provide evidence of your students' ability to use the scientific practices?

6. Based on what you learned about your students' ideas related to floating and sinking, what advice or suggestions do you have for your colleagues and future teachers?

7. Based on what you learned about using the P-E-O technique, what advice or suggestions do you have for your colleagues and future teachers?

GOING FURTHER

1. Read and discuss the Teacher Notes for the "Solids and Holes" probe on pages 42–46 (Keeley, Eberle, and Tugel 2007). Pay particular attention to the Related Research and Suggestions for Instruction and Assessment sections.

2. Examine the P-O-E activities in Haysom, J., and M. Bowen. 2010. *Predict, observe, explain: Activities enhancing scientific understanding.* Arlington, VA: NSTA Press. How can these be used for formative assessment?

3. Read and discuss the article: Smithenry, D. and J. Kim. 2010. Beyond predictions. *Science and Children* 48 (2): 40–44.

4. The seminal article, "Teaching for Conceptual Change: Confronting Children's Ideas," illustrates how difficult it is for children to give up their strongly held ideas, even when confronted with evidence from observations (Watson and Konicek 1990). Read and discuss the article at: *www.exploratorium.edu/ifi/resources/teachingforconcept.html.*

5. Watch and discuss the NSTA-archived web seminar on using the *NGSS* practices in the elementary grades at *http://learningcenter.nsta.org/products/symposia_seminars/NGSS/webseminar16.aspx.*

Chapter 23
Labeling Versus Explaining

In the elementary grades, the butterfly is a commonly used curricular context for children to learn about growth and development of organisms as they progress through their life cycle. *A Framework for K–12 Science Education's* life science core idea LS1.B, Growth and Development of Organisms, states that by the end of grade 5, students should know that plants and animals have unique and diverse life cycles that include being born (sprouting in plants), growing, developing into adults, reproducing, and eventually dying (NRC 2012, p. 146). The butterfly is an example of an organism with an interesting and unique cycle of birth, growth, and development that can readily be observed in the classroom or studied vicariously through picture books and video.

Typically, children learn that the egg is the first stage in the life cycle of a butterfly. A female butterfly deposits her eggs on the leaves of plants. The second stage begins when a caterpillar (larva) hatches from the egg. The caterpillar feeds on the plant and grows rapidly. When it stops eating and attaches itself to a leaf or other substrate, it begins the third stage in its life cycle, the chrysalis (pupa). The fourth state, the adult butterfly, begins when the butterfly emerges from the chrysalis. And thus the cycle continues as adult butterflies reproduce and lay new eggs. As children learn about these stages, they connect scientific words to their observations of life cycle phenomena such as *complete metamorphosis, egg, larva, caterpillar, pupa, chrysalis,* and *adult*. They draw labeled cycle diagrams that depict each of these stages with the correct scientific word and include descriptions, as well as detailed drawings, of the characteristics of each stage. Using the language of the butterfly life cycle and the labeled drawings in their science notebooks, children write about or engage in class discussions about the butterfly life cycle, such as the similarities and differences of each stage, data on the time it takes for the butterfly to complete each stage as well as its complete development from egg to adult, structures and functions of the organism at each stage, and observed behaviors. After completing the unit, a teacher might believe the instructional opportunities provided for the students developed their conceptual understanding of a life cycle and how the butterfly's life cycle differs from other organisms.

Vygotsky (1986), a well-known cognitive psychologist, described how children's understanding of concepts is based on the appearance and characteristics of the phenomena they experience in everyday situations and events. Furthermore, Vygotsky described how language becomes the tool through which children further develop their understanding and that many ideas are transmitted to children through the language adults use. Using the "Chrysalis" formative assessment probe (Figure 23.1), let's examine how students' ideas about the chrysalis stage of the butterfly's life cycle can reflect Vygotsky's research and what the implications are for teaching and learning (Keeley 2011). A key feature of using formative assessment probes is that they help teachers see the link between research on learning and their own students' ideas. For example, some students may believe that things must be visibly eating or breathing to be considered living. One of the most common misconceptions children have is that something is not alive if it does not move. In studies of children's understanding of the continuity of life, 19% of children ages 10–14 believed that larvae changed into pupae, which are dead, and then they became butterflies (Driver et al. 1994). Additionally, the ideas students reveal through their responses may inform aspects of instruction that teachers were not initially aware of, and they can take action to modify their instructional approach to address "hidden" misconceptions that existed during the teaching and learning process.

How do students' responses to this probe connect to Vygotsky's research and reveal information about teaching and learning? How can results from this formative assessment be used to modify instruction? Throughout the butterfly unit a teacher might conceptualize: egg (living)–larva (living)–chrysalis (living)–adult (living); but, a student interprets this as: egg (living but some students also see nonliving)–larva (living)–chrysalis (dead)–adult (living). What the teacher sees and conveys is not always what the student interprets. When students experience observations of the chrysalis stage during their outdoor encounters with the natural world or in the classroom, the chrysalis may not appear living to them based on the criteria they use to define living (as described in research). In addition, phrases carelessly used by adults, such as "new life emerges," contribute to a students' naive idea that a living butterfly can arise from a dead chrysalis. Erroneous ideas about the chrysalis stage are further compounded when the emphasis is on developing the scientific terminology for labeling and talking about the stages without connecting the words to scientific explanations. In a study of first graders' ways of seeing and talking about insect life cycles, Shepardson (1997) describes how language used with students must be viewed as a system for explaining

FIGURE 23.1.
The Chrysalis Probe

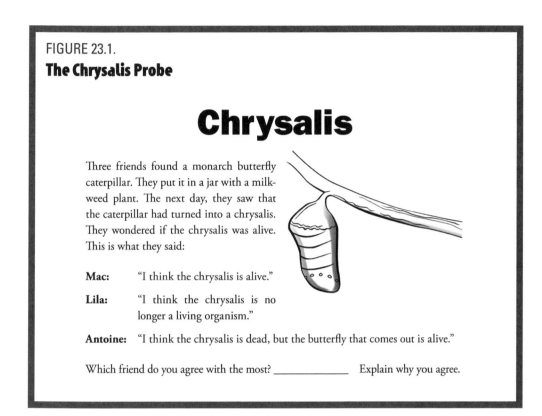

Chrysalis

Three friends found a monarch butterfly caterpillar. They put it in a jar with a milkweed plant. The next day, they saw that the caterpillar had turned into a chrysalis. They wondered if the chrysalis was alive. This is what they said:

Mac: "I think the chrysalis is alive."

Lila: "I think the chrysalis is no longer a living organism."

Antoine: "I think the chrysalis is dead, but the butterfly that comes out is alive."

Which friend do you agree with the most? _____ Explain why you agree.

and interpreting, versus labeling, so that children have the appropriate tools for seeing, acting, talking, and thinking about living phenomena such as life cycles.

What does this mean for assessment-informed instruction (formative assessment)? Place less emphasis on the terminology used for labeling purposes and more emphasis on using the terminology in developing explanations of what is happening during each stage of development. The scientific practice of constructing explanations in this context should involve students in explaining the phenomena of the metamorphosis of a butterfly as it goes through the stages of its life cycle and how each stage is a different stage in the *living* organism's growth and development. What evidence exists at each stage to support the idea that the organism is alive during each stage? Furthermore, if movement is used as the sole criteria students use to explain why the chrysalis is not alive, this indicates to the teacher the need to connect other characteristics of life to the chrysalis stage as well as provide an opportunity for students to observe how a chrysalis reacts to touch with a twitching motion.

The marvel of watching an adult butterfly emerge from a chrysalis is a never-ending delight for adults and children. Through the use of this probe, that delight is further enhanced by ensuring that children also develop the understanding and appreciation of how life changes as it transforms within that delicate structure.

REFERENCES

Driver, R., A. Squires, P. Rushworth, and V. Wood-Robinson. 1994. *Making sense of secondary science: Research into children's ideas.* London: RoutledgeFalmer.

Keeley, P. 2011. *Uncovering student ideas in life science, volume 1: 25 new formative assessment probes.* Arlington, VA: NSTA Press.

National Research Council (NRC). 2012. *A framework for K–12 science education: Practices, crosscutting concepts, and core ideas.* Washington DC: National Academies Press.

Shepardson, D. 1997. Of butterflies and beetles: First graders' ways of seeing and talking about insect life cycles. *Journal of Research in Science Teaching* (34) 9: 873–899.

Vygotsky, L. S. 1986. *Thought and language.* Cambridge, MA: MIT Press.

NSTA CONNECTION

Download the "Chrysalis" probe at *www.nsta.org/SC1302*. Read the entire chapter and the introduction to *Uncovering Student Ideas in Science, Volume 1: 25 Formative Assessment Probes* in the NSTA Science Store (*http://bit.ly/vaihtZ*).

● ●

Labeling Versus Explaining
Reflection and Study Guide

QUESTIONS TO THINK ABOUT AFTER YOU READ THIS CHAPTER

1. What should elementary students know about life cycles of organisms? What are some of the typical instructional experiences elementary students have to learn about life cycles? What are some ways elementary students typically demonstrate what they learned about life cycles?

2. What do you think Vygotsky meant by language being the tool through which children further develop their understanding and that many ideas are transmitted to children through the language adults use? How does the use of the "Chrysalis" probe illustrate this?

3. How do the formative assessment probes in the *Uncovering Student Ideas in Science* series help teachers link the research on learning to their own students' ideas?

4. What are some of the commonly held misconception children have about living things that may affect how they think about the chrysalis stage of the life cycle?

5. What do you think is meant by children having hidden misconceptions? How are the probes designed to reveal hidden misconceptions that may not surface when using your instructional materials?

6. What are some examples of "careless" language adults use that may cause or reinforce misconceptions?

7. Does labeling a diagram with the terminology used to describe a life cycle imply that a student has a conceptual understanding of life cycles? What is an alternative way for students to use the terminology associated with life cycles?

8. *A Framework for K–12 Science Education* and *The Next Generation Science Standards* emphasize scientific and engineering practices as one of the three dimensions of learning science. One of these practices is constructing explanations in science. How is this practice important for formative assessment purposes? (For more information about scientific and engineering practices visit NSTA's *NGSS* portal at *http://ngss.nsta.org*.)

PUTTING FORMATIVE ASSESSMENT INTO PRACTICE

1. What did you learn about your students' ideas after using this probe? Were you surprised by any of their answer choices and explanations?

2. Do you think your students could correctly label a diagram of the life cycle of a butterfly and still answer this probe incorrectly? If so, what does that tell you about the use of labeled diagrams as evidence of student learning?

3. As you analyze your students' explanations, did you find any evidence of how language may have affected their thinking? What insight did you gain into where your students' partial understanding or misunderstanding may have originated?

4. What instructional decisions did you make after analyzing your students' responses?

5. What feedback would you give your students on their ability to construct a scientific explanation? What suggestions do you have for improving the quality of their explanations?

6. Based on what you learned about your students' ideas related to life cycles, what advice or suggestions do you have for your colleagues and future teachers?

7. Based on what you learned about labeling versus explaining, what advice or suggestions do you have for your colleagues and future teachers?

GOING FURTHER

1. Read and discuss the Teacher Notes for the "Chrysalis" probe on pages 42–46 (Keeley, Eberle, and Tugel 2007). Pay particular attention to the Related Research and Suggestions for Instruction and Assessment sections.

2. Read and discuss the article about pre- and postassessment of plant life cycles: Schussler, E., and J. Winslow. 2007. Drawing on students' knowledge. *Science and Children* 44 (4): 40–44.

3. Read and discuss the Perspectives article: Gagnon, M., and S. Abell. 2008. *Science and Children* 45 (4): 60–61.

4. Watch and discuss the NSTA-archived web seminar on the *NGSS* scientific practice of constructing explanations at: *http://learningcenter.nsta.org/products/symposia_seminars/NGSS/webseminar10.aspx*

Chapter 24
When Equipment Gets in the Way

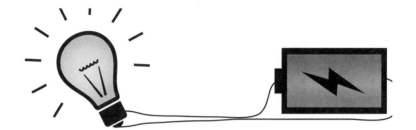

Electrical circuit units have been a mainstay in grades 3–5 science curricula for decades. Many classrooms use commercial elementary science kits that come complete with battery holders, screw-in lightbulb holders, bulbs, wire clips, switches, and wires to investigate electric circuits. With engineering gaining more prominence in the elementary classroom, students use a variety of electrical circuitry materials or kits to solve a problem by building devices that transform chemical energy from a battery into electrical energy that moves through a circuit to produce sound, light, or motion. These activities address several of *A Framework for K–12 Science Education's* scientific and engineering practices as well as core idea PS3.B: Conservation of Energy and Energy Transfers, which states that by the end of grade 5 students will know that energy can be transferred from place to place by electric currents, which can then be used locally to produce motion, sound, heat, or light (NRC 2012).

Despite these early circuitry activities, numerous studies have shown that students of all ages, including adults, have difficulty lighting a lightbulb when given only a battery, wire, and a bulb. Even Harvard graduates, as seen in the *A Private Universe* video, had difficulty when presented with this task (Private Universe Project 1995). As I watched the video before developing this probe, I was struck by the graduates' frustration of not having something to screw the lightbulb into. One graduate expressed her belief that it was not possible to light the bulb without a socket. One can assume that these students had experiences constructing electrical circuits in their high school physics class and may have had "batteries and bulbs" experiences in elementary school as well. So, why is it so difficult for adults to perform this basic elementary task? What lessons can we learn from children as they respond to a formative assessment probe and investigate electrical circuits?

Using the Probe

The formative assessment probe "Batteries, Bulbs, and Wires" (Figure 24.1, p. 169) may be used to elicit students' initial ideas about how to light a lightbulb with a battery and wire (Keeley, Eberle, and Dorsey 2008). Students' responses to the probe, along with their drawings, reveal ideas about the pathway of electricity from the battery to the bulb that mirrors research findings on children's ideas about electric circuits involving a battery, wire, and bulb.

Students who choose one wire will often use a source-consumer model in which the battery gives something to the bulb. Their drawings often show the wire going from the top of the battery to the bulb (unipolar model). The bipolar model used by students who choose two wires shows each wire going from the end of the battery to the tip of the bulb. Students explain how the electricity goes from each end of the battery and through the wire attached to each end, to the bulb to make it light (Driver et al. 1994). As a formative assessment given before students have had the opportunity to learn about complete circuits, the data clearly shows the need to provide students with a battery, bulb, and wire and have them investigate. As the teacher watches how students struggle to light the bulb and probes further to have students explain their configurations, the teacher is thinking about next steps needed in order for students to construct meaning about what is happening in the battery-wire-bulb system. One of these steps involves guiding students through recognizing the architecture of the bulb that allows electricity to flow in a complete loop.

I have also seen examples of student work from grades 3–5 students who selected either two or three wires, but added a lightbulb socket and a battery holder to their diagram. Students who chose three wires often added a switch to their drawing, explaining the need for a switch to keep the circuit closed. These students had prior experiences constructing circuits from kits—either science kits in school or electronic kits they played with at home—and needed to have the equipment represented in order to construct their model of a circuit. These students' responses to the probe are very similar to the high school physics students seen on the *A Private Universe* video. The physics students in the video used circuit boards to investigate circuits. They could construct different circuits to light a bulb with their circuit boards that had the power pack and bulb sockets built into it, but were unable to do so when given the materials (battery, bulb, wire) not embedded in the circuit board. What does this probe indicate to the teacher when students select two or three wires and draw pictures that show the wires going into the socket from two ends of a battery clip (and sometimes include a light switch) but are unable to figure out that a bulb can be lit using only one wire?

Responses from probes such as the ones described above can point out how scientific equipment sometimes gets in the way of building understanding. While it is nice to

FIGURE 24.1.

The Batteries, Bulbs, and Wires Probe

Batteries, Bulbs, and Wires

Kirsten has a battery and a small bulb. She wonders how many strips of wire she will need to connect the battery and the bulb so that the bulb will light. What is the *smallest number* of wire strips Kirsten needs to make the bulb light up?

A One strip of wire

B Two strips of wire

C Three strips of wire

D Four strips of wire

Explain your thinking about how to light the bulb. Draw a picture to support your explanation.

Picture:

_____

_____

_____

_____

_____

_____

have real circuitry materials to use in the classroom, the learning can get lost in the equipment that does the "learning" for them. By connecting wires to the bulb socket, students never really understand how the current flows through a lightbulb. By having wires that come off both ends of the battery holder, students never have to contemplate how electricity flows from the battery. The formative lesson that can be learned by teachers using this probe who discover similar conceptual gaps among students who have used circuitry materials is to start simple before adding equipment that takes away some of the discovery of how things work. The Private Universe Project suggested the following: Start with one wire. When feedback indicates that students understand how to build a circuit with only one wire, a battery, and a bulb, give them two wires. When they understand how to build a circuit with two wires, a bulb, and a battery, give them the bulb socket and the battery holders to construct a circuit. Start simple and build up from there.

As you use this formative assessment probe or similar probes in which students use scientific equipment to investigate questions and phenomena, look for evidence of materials getting in the way of conceptual learning. This is true of technology as well as other types of scientific equipment we use with students. Remember, the learning is not in the materials, it is in the sense making. Formative assessment probes can help teachers see when learning gets lost in the equipment and redirect to using simple materials that allow students to construct meaning first.

REFERENCES

Driver, R., A. Squires, P. Rushworth, and V. Wood-Robinson. 1994. *Making sense of secondary science: Research into children's ideas.* New York: RoutledgeFalmer.

Keeley, P., F. Eberle, and C. Dorsey. 2008. *Uncovering student ideas in science, volume 3: Another 25 formative assessment probes.* Arlington, VA: NSTA Press.

National Research Council (NRC). 2012. *A framework for K–12 science education: Practices, crosscutting concepts, and core ideas.* Washington, DC: National Academies Press.

Private Universe Project. 1995. *A Private Universe Teacher Workshop Series.* [Videotape] South Burlington, VT: The Annenberg/CPB Math and Science Collection.

NSTA CONNECTION

Download the "Batteries, Bulbs, and Wires" probe at *www.nsta.org/SC1303*. Read the entire chapter and the introduction to *Uncovering Student Ideas in Science, Volume 1: 25 Formative Assessment Probes* in the NSTA Science Store (*http://bit.ly/vaihtZ*).

When Equipment Gets in the Way
Reflection and Study Guide

QUESTIONS TO THINK ABOUT AFTER YOU READ THIS CHAPTER

1. With engineering gaining more prominence in the science curriculum, why is it important for elementary students to be able to construct and understand simple electrical circuits?

2. Most adults were taught basic electrical circuit ideas somewhere in their K–12 education. Many even constructed simple circuits with a battery, bulb, and wire. So, why do you think many adults have difficulty demonstrating how to light a bulb with a battery and wire?

3. How is the formative assessment probe, "Batteries, Bulbs, and Wires" different from other elementary grades' electrical circuit assessments or activities you have used or are familiar with? What does this probe uncover that other questions or tasks might miss?

4. Describe some of the research-identified conceptual models students use to explain how electricity flows in a simple circuit? Why do you think these ideas make sense to students, even after they have constructed circuits?

5. How does this probe reveal the need to help students recognize the structure of a lightbulb in order to explain how a simple circuit works?

6. What does the chapter mean by equipment getting in the way of learning science by "doing the learning" for students? Can you think of other examples of how equipment and hands-on materials used in science may inhibit conceptual understanding?

7. What does this chapter suggest as an instructional strategy that addresses the need for students to "start simple" when investigating simple circuits?

8. What does the chapter mean by "the learning is not in the materials, it is in the sense-making"? *A Framework for K–12 Science Education* and *The Next Generation Science Standards* emphasize connecting scientific and engineering practices to the content students are learning. What important message does this quote send to ensure students that connections between the content and practices of science will support conceptual understanding? (For more information about scientific and engineering practices visit NSTA's *NGSS* portal at: *http://ngss.nsta.org*.)

PUTTING FORMATIVE ASSESSMENT INTO PRACTICE

1. What did you learn about your students' ideas related to electrical circuits after using this probe? Were you surprised by any of their answer choices and explanations?

2. Did any of your students choose one wire yet their explanation and drawing revealed misconceptions about circuits? Why is it important to examine not only the answer students choose, but also their explanations?

3. How did your students' drawings provide you with greater insight into their ideas?

4. What instructional decisions did you make after analyzing your students' responses?

5. Did your students' responses indicate any evidence that equipment got in their way? If so, what changes will you make to your instructional materials?

6. Based on what you learned about your students' ideas related to simple circuits, what advice or suggestions do you have for your colleagues and future teachers?

GOING FURTHER

1. Read and discuss the Teacher Notes for the "Batteries, Bulbs, and Wires" probe on pages 58–62 (Keeley, Eberle, and Dorsey 2008). Pay particular attention to the Related Research and Suggestions for Instruction and Assessment sections.

2. *Uncovering Student Ideas in Physical Science, Volume 2: 39 New Electricity and Magnetism Formative Assessment Probes* (Keeley and Harrington 2014) has additional information in the introduction chapter on students' ideas about electrical circuits as well as an extensive collection of electrical circuit probes.

3. The Private Universe Project in Science has a series of online streaming videos that address students' ideas in science. Watch and discuss video workshop #3 "Hands-On Minds-On Learning" to learn more about students' commonly held misconceptions about electrical circuits that may be related to the experiences they had with circuitry equipment. This series can be accessed at *www.learner.org/ resources/series29.html* (Harvard-Smithsonian Center for Astrophysics 1995).

4. Read and discuss the article: Concannon, J., P. Brown, and E. Pareja. 2007. Making the connection: Addressing students' misconceptions of circuits. *Science Scope* 30 (3): 10–15.

5. Read and discuss the article: Hodgson-Drysdale, T., and E. Ballard. 2011. Explaining electrical circuits. *Science and Children* 48 (8): 37–41.

Chapter 25

Is It a Rock?
Continuous Formative Assessment

When elementary teachers ask, "When and how often should I use formative assessment?" my answer is "continuously, throughout instruction." Formative assessment can be used prior to a lesson for the purpose of eliciting students' preconceptions. It can also be used throughout the instructional cycle as students explore their ideas and develop and refine new conceptual understandings. Formative assessment is also used at the end of a sequence of instruction to provide an opportunity for students to refine their thinking and reflect back on how their ideas have changed. This month's column will show how the formative assessment probe "Is It a Rock?" (Keeley, Eberle, and Tugel 2007; Figure 25.1, p. 174) can be combined with a formative assessment classroom technique (FACT), the group Frayer Model (Keeley 2008), to continuously inform teachers of their students' progress toward meeting the intended learning goals and engage students in rethinking their ideas as new information is assimilated.

The following describes how the probe and this FACT can be used together throughout a learning cycle of continuous assessment and instruction to both promote learning and inform instruction.

The formative assessment probe, "Is It a Rock? (Version 2)" connects two related elementary core ideas from *A Framework for K–12 Science Education* (NRC 2012). In Physical Science Core Idea PS1, elementary students develop the idea that matter can be described and classified by its observable properties, by its uses, and by whether it occurs naturally or is manufactured. In Earth Science core idea ESS2, students develop the idea that the geosphere, one part of the Earth system, is made up of rock, soil, and sediments. The probe combines these two ideas to help teachers examine how their students use the concepts of properties of matter, natural versus human-made materials, and the geologic origin of a material to distinguish between rocks and materials that are "rocklike."

The Frayer Model, developed by Dorothy Frayer and her colleagues at the University of Wisconsin, was originally designed as a strategy to support concept mastery (Frayer, Frederick, and Klausmeier 1969). The strategy is frequently used in literacy to support vocabulary development in the content areas. As a formative assessment classroom technique (FACT), it activates students' thinking about a concept and can be used to assess conceptual understanding.

FIGURE 25.1.

The Is It a Rock? Probe

Is It a Rock? (Version 2)

What is a rock? How do you decide if something is a rock?
Put an X next to the things that you think are rocks.

____ cement block ____ piece of clay pot ____ coal

____ dried mud ____ coral ____ brick

____ hardened lava ____ limestone ____ a gravestone

____ asphalt (road tar) ____ iron ore ____ marble statue

____ glass ____ concrete ____ granite

Explain your thinking. What "rule" or reasoning did you use to decide if something is a rock?

Typically, students individually complete a Frayer Model worksheet by filling in the definition of a targeted concept (in their own words), characteristics of the concept, and examples and non-examples of the concept. However, the Frayer Model can also be used as an interactive strategy where students work collaboratively in small groups of three or four to come up with a collective answer to each of the sections of the Frayer Model. Instead of individual worksheets, prepare a classroom chart of the Frayer Model and give each group four different-color sticky notes, color-coded for each section of the chart. Each group discusses their ideas and comes up with a collective response for each section. Each group's responses are recorded on the notes and placed on the chart to share with the class (Figure 25.2).

Elicitation

In this phase of a learning cycle, the "Is It a Rock?" probe is used to activate students' thinking and draw out prior knowledge of rocks students bring to their learning. Introduce the probe and go over each item on the list, making sure students are familiar with each object. You may show an actual object from the list, such as a piece of granite, or pictures of the objects on the list, such as a marble statue, to make sure students have some familiarity with each object before responding to the probe. Students then answer the probe individually in writing.

Exploration and Discovery

In this phase, students form small groups

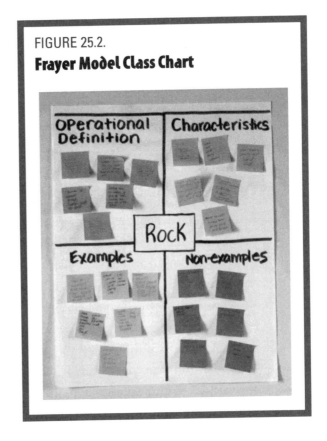

FIGURE 25.2.
Frayer Model Class Chart

to share their individual ideas with their peers and seek to discover new knowledge through the small-group discussion (science talk). Each student in the group explains his or her response to the probe, seeking feedback from the group, and evaluates the ideas of others in the group as they share their ideas. Then, introduce the group Frayer Model FACT, explaining how each group is to come up with a consensus response for the targeted concept of "rock" for each section of the chart. For the Examples and Non-Examples sections, instruct the students to use the objects listed on the "Is It a Rock?" probe.

After each group posts their responses on the chart and examines the sticky notes from other groups, summarize the chart for the class and point out similarities and differences in their ideas, taking care at this point not to pass judgment on whether students' ideas are right or wrong. Explain how the class will explore these ideas further and revisit the same probe again after students gather more information. Then, use the data from the group Frayer Model to formatively design a lesson or set of lessons that will move the class toward a scientific understanding of a rock as a natural material formed in the Earth through geologic processes. By examining the groups' responses, you may identify commonly held difficulties in distinguishing between things that are hard or made by humans out of earth materials and actual rocks made through geologic processes. Other difficulties, such as whether a hard, naturally formed material like coral is considered a rock, might also surface. All of these initial ideas are taken into account as you design lessons that will help the class construct new knowledge and understanding about what materials or objects are considered to be rocks.

Concept Development and Transfer

After students have had an opportunity to further explore their ideas, be confronted with new scientific information that challenges their initial ideas, and revise their thinking, have the students respond again to the "Is It a Rock?" probe. Give each student an opportunity to revise their original answers based on what they now know. Have students share their new thinking in their small groups and complete a new sticky note, which is placed over their first note to show how each group's ideas have changed (or in some cases, stayed the same). Facilitate a discussion with the whole class, giving each small group an opportunity to share and defend their ideas. At this point, address any difficulties or misunderstandings the students may still have. Now you are ready to guide the whole class in creating a class Frayer Model. Remove the sticky notes and, together with the class, fill in the Frayer Model to represent the class's scientific understanding of the concept of a rock.

Now that the students have developed a formal scientific understanding of what a rock is and how it differs from "rocklike" materials, add another assessment opportunity to see if students can transfer what they learned to new examples. List five new objects: block of ice, shell fossil, floor tile, quartz, and piece of pottery, and pass a sample of each around for the students to examine. Ask them to meet in their small groups and decide which ones are examples of rocks and which are non-examples, using the definition and characteristics on the whole-class Frayer Model chart. As each group shares their thinking, assess how well the class is able to transfer their new knowledge to categorizing new objects.

Reflection

Reflection provides a metacognitive opportunity for students to recognize how their thinking has changed as a result of their instructional experiences. Once again, ask students to examine their initial response to the probe and the class chart and think about how their ideas have changed up to this point. Each student then completes a reflection exit ticket by filling in the blanks: I used to think _____, but now I know _____. I know this because _____. Collect the reflections and examine them for evidence of conceptual change and indicators of the lessons' effectiveness.

This is just one example of the continuous use of the same formative assessment probe combined with a FACT to encourage group discussion, sharing, evaluation of ideas during different phases of instruction and learning, and reflection. The next time you select a probe and a FACT to use with that probe, consider how you can optimize their use throughout your sequence of instruction.

REFERENCES

Frayer, D., W. Frederick, and H. Klausmeier. 1969. *A schema for testing the level of concept mastery.* Madison, WI: Wisconsin Center for Education.

Keeley, P. 2008. *Science formative assessment: 75 practical strategies for linking assessment, instruction, and learning.* Thousand Oaks, CA: Corwin Press.

Keeley, P., F. Eberle, and J. Tugel. 2007. *Uncovering student ideas in science, volume 2: 25 more formative assessment probes.* Arlington, VA: NSTA Press.

National Research Council (NRC). 2012. *A framework for K–12 science education: Practices, crosscutting concepts, and core ideas.* Washington, DC: National Academies Press.

NSTA CONNECTION

Download the "Is It a Rock?" probe at *www.nsta.org/SC1304*. Read selections from the entire introduction to *Uncovering Student Ideas in Science* series at *www.nsta.org/publications/press/uncovering.aspx*.

Is It a Rock? Continuous Formative Assessment Reflection and Study Guide

QUESTIONS TO THINK ABOUT AFTER YOU READ THIS CHAPTER

1. What is continuous assessment and how does it fit into an instructional cycle?

2. What do you think is the purpose of the "Is It a Rock?" formative assessment probe? What ideas is this probe designed to elicit?

3. The Frayer Model is most known as a content literacy strategy. Have you used the Frayer Model? How can the Frayer Model be used for formative assessment in science?

4. How is the Frayer Model used interactively in a small-group format? What is the advantage to using the strategy in this way?

5. How did this chapter help you understand how formative assessment is used throughout the stages of an instructional cycle?

6. Formative assessment and instruction are described as being like two sides of the same coin. They are inextricably linked and one cannot be separated from the other. How did the examples in this chapter illustrate the link between instruction and assessment?

7. How does the use of the "Is It a Rock?" probe during the elicitation and exploration phase of instruction differ from how it is used in the concept development and transfer phase of instruction? How is it used formatively in each phase?

8. Why is it important to make time for student reflection? What are some strategies you can use to encourage students to reflect back on how their ideas have changed?

PUTTING FORMATIVE ASSESSMENT INTO PRACTICE

1. What did you learn about your students' ideas related to rocks as natural resources after using this probe? Were you surprised by any of their answer choices and explanations?

2. Which objects on the justified list did your students grapple with the most? Why do you think those objects challenged their thinking more than some of the other objects?

3. Did you use the Group Frayer Model? How did your students respond to using this strategy? Would you use this strategy with other probes? What other interactive strategies could be used with this probe?

4. Describe how formative assessment was embedded in the stages of the instructional model you used.

5. Based on what you learned about your students' ideas related to rocks, what advice or suggestions do you have for your colleagues and future teachers?

GOING FURTHER

1. Read and discuss the Teacher Notes for the "Is It a Rock?" probe on pages 158–162 (Keeley, Eberle, and Tugel 2007). Pay particular attention to the Related Research and Suggestions for Instruction and Assessment sections.

2. Read about and try out other formative assessment classroom techniques (FACTs) that can be used with the formative assessment probes at different stages

in an instructional cycle in: Keeley, P. 2008. *Science formative assessment: 75 strategies for linking assessment, instruction, and learning* and Keeley, P. 2014. *Science formative assessment: 50 more practical strategies for linking science, instruction, and assessment*. Thousand Oaks, CA: Corwin Press.

3. Read and discuss an article that describes another way to use the "Is It a Rock?" probe with the 5E instructional model: Brkich, K. 2012. Science shorts: Is concrete a rock? *Science and Children* 50 (4): 80–82.

4. Watch and discuss the archived web seminar on formative assessment at: *http://learningcenter.nsta.org/products/symposia_seminars/NLC/webseminarVII.aspx.*

Chapter 26

Is It a Solid?
Claim Cards and Argumentation

A *Framework for K–12 Science Education's* disciplinary core idea PS1.A states that students should know by the end of grade 2 that different kinds of matter exist and many of them can be solid or liquid, depending on temperature (NRC 2012). By the end of grade 8, they describe solids, liquids, and gases by the arrangement and motion of their molecules. But what about "in between" ideas?

In the elementary grades, students are typically taught to define solids and liquids using macroscopic properties. Solids generally keep their shape and have a definite volume. Liquids have a definite volume but can take the shape of their container. Sometimes this definition is expanded to include that liquids can be poured. This definition using shape and pouring can be problematic when tiny parts of solid materials, such as powders or granules, are involved. In addition, the everyday use of the word *solid* implies something that is hard and not soft or "airy." The assessment probe, "Is It a Solid?" (Figure 26.1, p. 183; Keeley, Eberle, and Dorsey 2008) can be used to elicit elementary students' ideas about solids and the macroscopic properties they use to decide whether a material is a solid. It can reveal whether students have developed misinterpretations based on early definitions of solids and liquids and the familiar use of these words. Furthermore, when combined with the formative assessment strategy known as *claim cards*, the probe also provides insight into how elementary students engage in the scientific practice of argumentation.

Making Claims

Using claim cards involves printing each of the objects or materials that make up the answer choices onto separate cards. Each student is then given a card. Students sit in a circle, each holding a card. They begin with a whip-around the circle, each student sharing the item on their card and stating their claim as to whether they think their item is or is not a solid. At this point there is no discussion of the validity of the claims. Going around the circle, each student states their claim as others listen and thinks about which claims they agree with and which ones they might challenge. After each student has shared his or her claim, the teacher asks for a volunteer to begin the science talk by restating his or her claim and justifying his or her thinking. She reminds the class of the norms for argumentation in science and that we respectfully argue in science to seek understanding, not win the argument. The following is a snapshot of the science talk that ensues in a fifth-grade classroom:

- **Kara:** "My claim is that wood is a solid. My evidence is that a block of wood keeps its shape, and if you measure the amount it stays the same."

- **Max:** "But you can change its shape by cutting it."

- **Kara:** "Yes, but whatever shape you cut it in, it holds that shape. It doesn't spread out or anything."

- **Freddie:** "Yeah, solids are hard, and wood is hard, so I agree with the claim."

- **Teacher:** "Does anyone want to add to or disagree with Kara's claim? Hearing no other ideas, let's for now put wood under solid, on our claims chart. Who would like to go next?"

- **Ivy:** "I have flour. My claim is flour is not a solid. It is a liquid. First I thought it might be a solid because it isn't wet like a liquid."

- **Teacher:** "Ivy, you said it isn't wet like a liquid, but you claim it is a liquid. Can you tell us why you think it is a liquid?"

- **Ivy:** "Well, it's like what we learned about liquids. You can pour them out and when you put them in something, they fill the shape of the thing you put them in like water does when you put it in a cup. You can pour it and the water fills out the cup. So I think flour does the same thing."

- **Teacher:** "Sharla, you want to add something?"

- **Sharla:** "I agree with Ivy. When me and my sisters make cookies, we pour the flour, and it's not hard like wood is."

- **Teacher:** "Who else has an idea to support or disagree with Ivy's claim?"

- **Hector:** "I'm not sure. It's kinda like water but not really, so I think it might be a solid."

- **Pete:** "But it doesn't keep the same shape like wood."

- **Teacher:** "We seem to have two different ideas about flour—some of you think it is a solid, some think it is a liquid, and others are not sure. Let's take a vote on where to put it on our claims chart. Remember, we will go back to our claims chart after our discussion, and you will have a chance to change your claims after you hear more arguments about whether the things on your cards are solids or not solids." Class vote indicates most think flour is a liquid, so the teacher lists flour under "not a solid" on the claims chart.

As each child shares his or her claim and justifies it with reasons for why it is or is not a solid, the teacher listens carefully for evidence of misconceptions or understanding about the properties of solids, liquids, and gases. She notes difficulties they

FIGURE 26.1.

The Is It a Solid? Probe

Is It a Solid?

What types of things are solid forms of matter? Put an X next to the things on the list that are solids.

_____ rock

_____ rubber band   _____ Styrofoam

_____ milk   _____ air   _____ ice

_____ feather   _____ flour   _____ wood

_____ cloth   _____ dust   _____ soil

_____ baby powder   _____ cooking oil   _____ melting wax

_____ sugar   _____ sponge   _____ salt

_____ foam-rubber ball   _____ iron nail   _____ cotton ball

Explain your thinking. What definition, rule, or reasoning did you use to decide whether something is a solid?

seem to have with powders, soft materials like the cotton ball, stretchy things like the rubber band, and things that are light or airy like the sponge and dust. She realizes that part of the definition they used to describe liquids—anything that fills the shape of its container—is posing problems for the students when they consider powders or granular substances.

As she examines the Related Research summaries in the Teacher Notes that accompany the probe (Keeley, Eberle, and Dorsey 2008), she finds that her fifth-grade students' ideas mirror the research on commonly held ideas students at that age may have. She realizes she has to design an experience using hand lenses for them to see that flour, baby powder, and sugar are made up of very tiny pieces of solid material. In the section of the Teacher Notes on Suggestions for Instruction and Assessment, she reads about a bridging analogy she might use with collections of large pieces like marbles in a jar.

She thinks she will use that as a demonstration and have her students connect it to their observations of the flour, baby powder, and sugar to see that these are really solids made up of a collection of individual solid pieces, and that some of these pieces can be very small. She also realizes she needs to give them multiple experiences to investigate solids that are soft and stretchy.

As the teacher reflects on the initial claims and argumentation session her students engaged in, she feels her students are really starting to grasp the practice of stating a claim and backing it up with evidence, even though their initial evidence may come from their naïve conceptions about matter. By revisiting the claims chart after they have had opportunities to be confronted with their claims and work out some of the disparities on the claims chart, she will be able to formatively assess the extent to which they can now draw upon evidence from their investigations and further research into characteristics of solids, liquids, and gases as they engage in argumentation.

This example shows how you can use formative assessment to uncover and—through carefully designed instruction—confront students' ideas about the macroscopic properties of solids before they move onto microscopic properties in middle school. Simultaneously, it provides an informal assessment window into students' ability to state claims and engage in argumentation. *Ready-Set-Science* states that "In spite of the importance of talk and argument in science and in the learning process in general, K–8 science classrooms are typically not rich with opportunities for students to engage in these more productive forms of communication" (Michaels, Shouse, and Schweingruber 2008, p. 89). There are many examples of assessment probes in the *Uncovering Student Ideas in Science* series you can use with the claim cards strategy to support the scientific practice of argumentation while developing deeper understanding of disciplinary core ideas.

REFERENCES

Keeley, P., F. Eberle, and C. Dorsey. 2008. *Uncovering student ideas in science, volume 3: Another 25 formative assessment probes*. Arlington, VA: NSTA Press.

Michaels, S., A. Shouse, and H. Schweingruber. 2008. *Ready, set, SCIENCE! Putting research to work in K–8 classrooms*. Washington, DC: National Academies Press.

National Research Council (NRC). 2012. *A framework for K–12 science education: Practices, crosscutting concepts, and core ideas*. Washington, DC: National Academies Press.

NSTA CONNECTION

Download the "Is It a Solid?" probe at *www.nsta.org/SC1307*.

Is It a Solid? Claim Cards and Argumentation Reflection and Study Guide

QUESTIONS TO THINK ABOUT AFTER YOU READ THIS CHAPTER

1. How does describing and classifying different forms of matter develop from grades K–2 to 3–5 to 6–8?

2. What do you think the chapter means by an "in between idea"?

3. The macroscopic definition of a solid as a form of matter that keeps its own shape is an early "stepping-stone" definition used with young children before they define solids based on the arrangement and motion of the particles. How can definitions used in earlier grades as stepping-stones to more complex definitions sometimes lead to misunderstanding? Does this mean we should not use stepping-stone definitions? How can formative assessment be used to check for misunderstandings when students apply a stepping-stone definition?

4. Describe how the claim cards technique was used with the "Is It a Solid?" probe. How would you use this technique?

5. What did you learn about the students' ideas in the short transcript provided in the chapter? How would you describe the teachers' role during the science talk?

6. Are you familiar with talk moves? What are talk moves and which ones did the teacher use during the science talk?

7. Describe the formative decision the teacher made based on listening to her students' share and discuss their ideas. What was the teacher thinking? How did she connect her knowledge of the research on learning related to solids to the evidence she gathered during the science talk? How did her instructional decision address her students' misunderstandings?

8. Do you think the teacher would have provided that instructional opportunity if she had not used a formative assessment probe to check on her students' descriptions of solids and liquids? How does this classroom example illustrate the importance of formative assessment to check on how students think about and apply a concept or definition?

9. Describe how the students used claims and evidence to engage in argumentation. How can use of a formative assessment probe combined with formative assessment classroom techniques (FACTs) such as claim cards, support the scientific practice of argumentation? (For more information about the scientific and engineering practices visit NSTA's *NGSS* portal at *http://ngss.nsta.org*.)

10. The chapter ends with a quote from *Ready, Set, Science!* that states that "In spite of the importance of talk and argument in science and in the learning process in general, K–8 science classrooms are typically not rich with opportunities for students to engage in these more productive forms of communication" (Michaels, Shouse, and Schweingruber 2008, p. 89). Do you agree with this quote? What can you do at the classroom and school level to support opportunities for students to engage in rich, productive science talk?

PUTTING FORMATIVE ASSESSMENT INTO PRACTICE

1. How does your curriculum address describing and classifying states of matter? How do you define solids and liquids at your grade level?

2. What did you learn about your students' ideas by examining their responses to the probe? Were you surprised by any of their responses?

3. Formative assessment drives next steps for instruction. It also helps teachers re-vise how they may teach an instructional unit in the future. Describe changes you might make to your instructional unit the next time you teach it, based on what you learned from your students.

4. Did you use claim cards or other strategies to engage students in talk and argu-ment? If so, what did you learn from listening to your students? How did the strategy work in your classroom? What did you learn about your students' abil-ity to use claims and evidence to support their arguments?

5. Did you provide feedback to your students to improve their ability to engage in productive talk? If so, what feedback did you provide?

6. What will you do (or did you do) to address your students' misconceptions about solids and liquids? Describe how your next steps are driven by your stu-dents' ideas.

7. Based on what you learned from using this probe with your students, what sug-gestions do you have for your colleagues and future teachers?

Going Further

1. Read and discuss the Teacher Notes for the "Is It a Solid?" probe (Keeley, Eberle, and Dorsey 2008, pp. 26–31). Pay particular attention to the Related Research and Suggestions for Instruction and Assessment sections.

2. Read and discuss the section on engaging in argument from evidence on pages 71–74 in the *A Framework for K–12 Science Education* (NRC 2012) or online at

www.nap.edu/openbook.php?record_id=13165&page=73. Examine the progression to learn how young students begin to engage in this practice.

3. The concept of different types of matter is included in the *NGSS* and *A Framework for K–12 Science Education* under PS1.A: Structure and Properties of Matter. Read the section on pages 106–109 or online at *www.nap.edu/openbook. php?record_id=13165&page=106.*

4. *Ready, Set, Science!* (Michaels, Shouse, and Schweingruber 2008) describes productive talk in chapter 5, Making Thinking Visible: Talk and Argument. This resource can be read online at *www.nap.edu/catalog.php?record_id=11882.*

5. Explore other formative assessment strategies that support talk and argument in *Science Formative Assessment: 50 More Practical Strategies that Link Assessment, Instruction, and Learning* (Keeley 2014).

6. Read and discuss the introduction to *Uncovering Student Ideas in Primary Science* (Keeley 2013). The introduction includes a section on using talk moves with children and supporting the *Common Core* literacy capacities of speaking and listening. This chapter is available as a free download through NSTA Press at *www. nsta.org/store/product_detail.aspx?id=10.2505/9781936959518* (Click on the link to "Is It Made of Parts?"

7. Read and discuss the article Ross, D., D. Fisher, and N. Frey. 2009. The art of argumentation. *Science and Children* 47 (3): 28–31. The article discusses language frames that can be used to support argumentation.

8. Read and discuss the article, Palmieri, A., A. Cole, S. deLisle, E. Erickson, and J. Janes. 2008. What's the matter with teaching children about matter? *Science and Children* 46 (4): 20–23. The article describes how to use children's ideas about matter to inform instructional planning, including ideas about solids.

9. Watch and discuss the NSTA-archived *NGSS* webinar on the scientific and engineering practice of engaging in argument from evidence at *http://learningcenter. nsta.org/products/symposia_seminars/NGSS/webseminar11.aspx.*

Chapter 27

When Is the Next Full Moon?
Using K–2 Concept Cartoons

Coincept cartoons, formative assessment tools that reveal students' preconceptions and probe for conceptual understanding (Naylor and Keogh 2000), have recently become popular in the United States, with teachers developing their own versions. "When Is the Next Full Moon?" (Figure 27.1, p. 191) is an elementary science probe with a concept cartoon format (Keeley 2013). This format provides an engaging, visual stimulus for eliciting students' ideas. It also serves as a starting point for their investigation. Characteristics of concept cartoons that make them effective formative assessment tools for younger students include:

- a visual format with minimal text
- characters talking with each other, each sharing a different idea
- familiar contexts or situations
- a "What are you thinking?" prompt to encourage students to share their ideas and explanations in a science talk format

Content Connections

The K–2 Earth and space science *Next Generation Science Standards* (*NGSS*) include the disciplinary core idea that patterns of the motion of the Sun, Moon, and stars in the sky can be observed, described, and predicted (NGSS Lead States 2013; see Internet Resource). "When Is the Next Full Moon?" is designed to address one component of this core idea—observing, describing, and predicting the monthly pattern of Moon phases. Let's explore how the concept cartoon format helps students develop this important understanding and prepares them to meet the *NGSS* grade 1 performance expectation 1-ESS1-1 Use observations of the Sun, Moon, and stars to describe patterns that can be predicted (NGSS Lead States 2013; see Internet Resources). Current state standards contain similar assessment expectations.

This probe may be used at the start of a unit on observing the pattern of Moon phases (ideally on the date when there is a full Moon). First, make sure students know what a full Moon is, and then present the cartoon. Working with a partner or in a small group, students decide which character they most agree with and why. The teacher then engages the whole class in science talk, listening carefully as pairs or small groups share the character they most agree with and explain their thinking, taking note of ideas students may have about Moon phases that can be used to inform subsequent instruction. The teacher then challenges the class to come up with a way the class could investigate to find out which cartoon character has the best idea about when a full Moon will appear again.

Through discussion, you should now have a record of understandings and misunderstandings students have about the pattern of Moon phases that can be revisited after the students have had an opportunity to observe, analyze their data, and revise or refine their initial claim and explanation. Consider whether some of the students have limited opportunity to see the night sky, and make note of times during the day the class can go outside to see the Moon or when students may have to use digital media to make their observations. As the class plans for and engages in an investigation to record and describe the daily changes to the shape of the Moon, they are using two of the scientific practices in the *NGSS*: (1) planning and carrying out investigations and (2) analyzing and interpreting data. These practices are further explicated for grade 1 as:

- Make observations (firsthand or from media) to collect data that can be used to make comparisons.
- Use observations (firsthand or from media) to describe patterns in the natural world in order to answer scientific questions (NGSS Lead States 2013, p. 14).

What's Next?

A critical feature of formative assessment involves giving students feedback. As students plan and carry out their Moon investigation, provide feedback to individuals and groups on their use of the scientific practices. This feedback is designed to move their learning forward so that they gain proficiency in the use of the practices to develop understanding of the scientific content.

After students have completed their investigation, the class comes together to analyze the data and discuss which probe character they now agree with and how the evidence from their observations support the character's idea they chose. They revisit some of the earlier alternative ideas shared during the preinvestigation discussion and talk about how their views have changed, sharing the evidence that helped them change their ideas. They decide whether observing one cycle is enough to support the class's theory that the Moon goes through a pattern of phases about every four weeks. They also raise additional questions, which often provide further insight into their thinking. To make sure that students can apply their knowledge using the data from their investigation to make predictions, ask new questions such as, "If there is a crescent Moon today, how

FIGURE 27.1.
The When Is the Next Full Moon? Probe

When Is the Next Full Moon?

What are you thinking?

long will it be before we see the same shape Moon again?" and "If there is a full Moon today, how many days will it take until we can see a crescent Moon?" All of the formative assessment data the teacher collects throughout the Moon phase investigation feeds into the summative assessment question, "Are my students ready to demonstrate that they can meet the Moon part of the performance expectation?" (1-ESS1-1 Use observations of the Sun, Moon, and stars to describe patterns that can be predicted.)

The concept cartoon format is particularly effective for formatively assessing students' readiness to meet this *NGSS* performance expectation as it provided a visual starting point for eliciting and sharing their initial ideas, launched them into an investigation that took their initial ideas into account, and provided an opportunity to revisit their initial ideas using evidence from data collected through the investigation and explanations developed through the sense making discussion. Throughout the full cycle of formative assessment, the concept cartoon provided an engaging way for students to learn that the pattern of Moon phases takes about four weeks and to use the scientific practices to develop that knowledge. It also provided a way for the teacher to gain access to students' ideas, including their misconceptions, and look for evidence of whether and how their ideas had changed or developed further as they progressed through their investigation.

As you start this new school year, consider developing your own concept cartoons or using ones that are already published to engage your students, provoke productive classroom discussion that uncovers their existing ideas, stimulate their scientific thinking, and launch them into purposeful investigations that address core ideas in science. In this way, a formative assessment probe becomes a valuable and deliberate part of the elementary science teaching and learning process.

REFERENCES

Keeley, P. 2013. *Uncovering student ideas in primary science, volume 1: 25 new formative assessment probes for grades K–2.* Arlington, VA: NSTA Press.

NGSS Lead States. 2013. *Next Generation Science Standards: For states, by states. www. nextgenscience.org/next-generation-science-standards.*

Naylor, S., and B. Keogh. 2000. *Concept cartoons in science education.* Cheshire, UK: Millgate House Publishers.

INTERNET RESOURCE

NGSS Table: 1-ESS1 Earth's Place in the Universe
www.nextgenscience.org/1ess1-earth-place-universe

NSTA CONNECTION

Download the "When Is the Next Full Moon?" probe at *www.nsta.org/SC1309.*

● ●

When Is the Next Full Moon?
Using K–2 Concept Cartoons
Reflection and Study Guide

QUESTIONS TO THINK ABOUT AFTER YOU READ THIS CHAPTER

1. How is the concept cartoon format different from other formats used in the *Uncovering Student Ideas series?* What do you think is the advantage to using concept cartoons over other formative assessment formats?

2. The idea that patterns in the night sky can be observed, described, and predicted is addressed in the *NGSS* and most state standards at the elementary level. How and why would you use the "When Is the Next Full Moon?" probe prior to students launching into moon observations?

3. Which *NGSS* scientific and engineering practices are supported through use of this probe?

4. What is the purpose of feedback and why is providing feedback important? How does the use of a formative assessment probe fit into a feedback cycle?

5. Why is it important for students to revisit their initial ideas after completing an investigation?

6. How can you assess whether students can apply the ideas they developed through an investigation to answer a different question?

7. How would you describe the link between formative and summative assessment? How can you use formative assessment probes and formative assessment classroom techniques (FACTs) to assess students' readiness to demonstrate their understanding of a learning target or performance expectation by taking them through a full cycle of assessment and instruction?

8. Describe how use of the "When Is the Next Full Moon" probe can prepare students for the *NGSS* grade 1 performance expectation 1-ESS-1 Use observations of the Sun, Moon, and stars to describe patterns that can be predicted.

9. Can you think of examples of content to use in developing your own concept cartoons and use them to formatively assess students throughout a cycle of assessment and instruction?

10. *A Framework for K–12 Science Education* and *The Next Generation Science Standards* emphasize crosscutting concepts as one of the three dimensions of science learning. How can the crosscutting concept of Patterns support students' understand-

ing of moon cycles when using this probe? (For more information about crosscutting concepts visit NSTA's *NGSS* portal at *http://ngss.nsta.org*.)

PUTTING FORMATIVE ASSESSMENT INTO PRACTICE

1. When and how does your curriculum or instructional materials address the pattern of Moon phases?

2. What did you learn about your students' ideas by examining their responses to the probe? Were you surprised by any of their responses?

3. Describe how you used this probe throughout a full cycle of assessment and instruction. What formative decisions did you make as a result of using this probe?

4. Was there an advantage to using the concept cartoon format? How did the cartoon format encourage and support your students' thinking?

5. Did you provide feedback to your students to improve their ability to engage in productive talk? If so, what feedback did you provide?

6. Based on what you learned from using this probe with your students, what suggestions do you have for your colleagues and future teachers?

GOING FURTHER

1. Read and discuss the Teacher Notes for the "When Is the Next Full Moon?" probe (Keeley 2013 pp. 114–116). Pay particular attention to the Related Research and Suggestions for Instruction and Assessment sections.

2. Read and discuss the section on the crosscutting concept of Patterns on pages 85–87 in *A Framework for K–12 Science Education* (NRC 2012) or online at *www.nap. edu/openbook.php?record_id=13165&page=85*.

3. Observing the pattern of moon phases is addressed in the *NGSS* and *A Framework for K–12 Science Education* under core idea ESS1-Earth's Place in the Universe. Read the sections on pages 173–176 or online at *www.nap.edu/openbook. php?record_id=13165&page=173*.

4. Learn more about concept cartoons and the research that underlies them by exploring the site of the original concept cartoons at *www.conceptcartoons.com*.

5. Read and discuss an article that describes how young children investigated the moon and how misconceptions were surfaced: Fitzsimmons, P., D. Leddy, L. Johnson, S. Biggam, and S. Locke. 2013. The moon challenge. *Science and Children* 50 (1): 36–41.

6. Watch and discuss the NSTA-archived *NGSS* webinar on the scientific and engineering practice of analyzing and interpreting data at: *http://learningcenter.nsta. org/products/symposia_seminars/NGSS/webseminar8.aspx*.

Chapter 28

Pendulums and Crooked Swings: Connecting Science and Engineering

The *Next Generation Science Standards* provide opportunities for students to experience the link between science and engineering. In the December 2011 issue of *Science and Children*, Rodger Bybee explains: "The relationship between science and engineering practices is one of complementarity. Given the inclusion of engineering in the science standards and an understanding of the difference in aims, the practices complement one another and should be mutually reinforcing in curricula and instruction" (p. 15).

In this month's column, we will explore this complementary nature of science and engineering in the elementary classroom. While engineering practices can be used effectively to help elementary students develop scientific knowledge, scientific knowledge gained through use of the scientific practices can also be used to solve engineering problems. The example used in this column will primarily address one aspect of this duality by focusing on how formative assessment can be used to assess students' readiness to define and carry out an engineering problem using knowledge gained through use of the scientific practices.

The Motion and stability: Forces and interactions performance expectation 3-PS2-2 states: "Make observations and/or measurements of an object's motion to provide evidence that a pattern can be used to predict future motion" (NGSS Lead States 2013, p. 23). The clarification statement that accompanies this performance expectation provides examples of motions that can be used, such as a child swinging on a swing (Note: the assessment boundary does not include terms like *period* and *frequency*). Since a swing is a type of pendulum, planning and carrying out a pendulum investigation provides an instructional opportunity for students to combine a disciplinary core idea, a scientific practice, and two crosscutting concepts. Furthermore, students can apply the results of their investigation to design a solution to an engineering problem (see Table 28.1, p. 197).

The formative assessment probe "The Swinging Pendulum" (see NSTA Connection) can be used to elicit students' initial ideas and launch into an investigation to discover what affects the number of swings a pendulum makes in a given time (Figure 28.1, p. 196; Keeley and Harrington 2010). There are two reasons to start an investigation with this probe: (1) If students correctly predict *B: shorten the string* and can support their prediction with an explanation, then why spend valuable class time carrying out an

FIGURE 28.1.
The Swinging Pendulum Probe

The Swinging Pendulum

Gusti made a pendulum by tying a string to a small bob. He pulled the bob back and counted the number of swings the pendulum made in 30 seconds. He wondered what he could do to increase the number of swings made by the pendulum. If Gusti can change only one thing to make the pendulum swing more times in 30 seconds, what should he do? Circle what you think will make the pendulum swing more times.

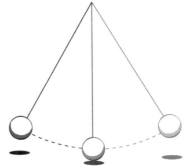

A Lengthen the string.

B Shorten the string.

C Change to a heavier bob.

D Change to a lighter bob.

E Pull the bob back farther.

F Don't pull the bob back as far.

G None of the above. All pendulums swing the same number of times.

Explain your thinking. What rule or reasoning did you use to select your answer?

investigation when students can already explain the cause and effects of changing the length of a string, changing the heaviness of the bob, or changing the release angle? If they already have this knowledge, then students are ready to use this knowledge to solve an engineering problem; (2) Committing to a prediction when students are not sure about the outcome creates a desire to gain new knowledge as students carry out the investigation. Let's see how this probe is formatively used to promote student learning and inform instruction that will help students gain the knowledge they will need to solve an engineering problem.

TABLE 28.1.

Science and Engineering Practices	Disciplinary Core Ideas	Crosscutting Concepts
Planning and Carrying Out Investigations • Plan and conduct an investigation collaboratively to produce data to serve as the basis for evidence, using fair tests in which variables are controlled and the number of trials considered. • Make observations and/or measurements to produce data to serve as the basis for evidence for an explanation of a phenomenon or test a design solution. **Constructing Explanations and Designing Solutions** • Apply scientific ideas to solve design problems.	**PS2.A: Forces and Motion** • The patterns of an object's motion in various situations can be observed and measured; when that past motion exhibits a regular pattern, future motion can be predicted from it.	**Patterns** • Patterns of change can be used to make predictions. (3-PS2-2) **Cause and Effect** • Cause and effect relationships are routinely identified.

Pendulums in Action

As students plan and carry out a scientific investigation of pendulums using simple materials such as string and washers, observe and listen closely for evidence that students can perform the practice of Planning and Carrying out Investigations. To what extent can they design a fair test by controlling each variable? How do they collect their data? Do they consider the number of trials? Do they use accurate techniques for measuring time, length of string, number of swings, and angle of release? If you observe that some students are struggling with the practices, provide constructive feedback to students to move their investigation and learning forward. For example, alert these students to the need to keep all the other variables the same when one is changed. If you observe that several groups are having a similar problem, stop and teach a minilesson or demonstrate a technique to the whole class, such as how to secure the string to a fixed pivot point rather than holding it in your hand, which may affect the swings. During a "scientists' meeting" where the class comes together to share its results and methods of investigating, provide feedback on how students carried out their investigation as they provide the evidence from data that supports their findings. Also, use probing questions to help students recognize the patterns in their data and cause-and-effect relationships.

To see if students can transfer their new learning about the motion of a pendulum, read The Crooked Swing story from *Yet More Everyday Science Mysteries* (Konicek-Moran 2011, p. 166). Transfer of learning (applying knowledge to new situations or contexts) is an important part of formative assessment that takes place after students have had

FIGURE 28.2.
The "Crooked" Swing

Taken from Yet More Everyday Science Mysteries (Konicek-Moran 2011)

an opportunity to develop scientific ideas. Engineering contexts can be an ideal vehicle to assess whether students can use their scientific knowledge to define an engineering problem.

In the story, children are presented with a problem. A neighbor hung a porch swing from a crooked branch of a large tree in his yard (Figure 28.2). The neighbor did his best to hang the swing so it was level with the ground. But lo and behold, when he and his wife sat on it, it swung crooked, rather than back and forth like a normal swing. The neighbor was perplexed and asked the children to help him figure out what the problem was and how to fix it. The students are tasked with finding a solution to the story.

This is a wonderful engineering problem for elementary students that uses scientific knowledge they gained from using the scientific practices and applies it to an engineering problem. First, students need to know something about pendulums in order to recognize the problem of the crooked swing. As groups grapple with the problem, listen to see whether students recognize that there are two pendulums—one shorter and one longer—holding up the swing. Based on what they learned about pendulums, do they recognize that one end will swing faster/slower than the other, resulting in a crooked path? Once students recognize that there are two pendulums of different lengths that swing differently, they are ready to engage in using engineering practices to come up with a way to fix the swing. Next, the story launches students into the design process: defining the problem, brainstorming possible solutions, considering constraints, developing and using a model to test their solutions, collecting and analyzing data on how their proposed solution works, refining it, and communicating it to others. Throughout the engineering activity, observe how students systematically use the design process and which parts of the process they may have difficulty with. Use a formative assessment observation sheet such as the one in Table 28.2 to record observations and provide feedback as you circulate among groups.

A Framework for K–12 Science Education states two implications related to the "Swinging Pendulum" probe and Crooked Swing story: "The first is that students should learn how scientific knowledge is acquired and how scientific explanations are developed. The second is that students should learn how science is utilized, in particular through the engineering design process, and they should come to appreciate the distinctions and

TABLE 28.2.

The Crooked Swing Problem Observation Sheet

Steps in the Engineering Design Process	Observations and Feedback
Identifying the problem	
Using scientific knowledge to define the problem	
Generating possible solutions (brainstorming ideas)	
Identifying constraints	
Selecting the best possible solution	
Constructing a prototype (model)	
Testing and evaluating the prototype (model)	
Improve the design	
Communicate the solution	
Other comments or questions:	

relationships between engineering, technology, and applications of science" (NRC 2012, p. 201). As a final reflection, after students have completed the pendulum investigation and engineering activity, ask them to write about how scientific ideas can be used to solve engineering design problems and how they think engineering design problems can help them learn science. This final reflective writing can be used formatively to see whether students also recognize the other aspect of the dual nature of engineering: how engineering problems can contribute to our knowledge about science.

REFERENCES

Bybee, R. 2011. Scientific and engineering practices in K–12 classrooms: Understanding a framework for K–12 science education. *Science and Children* 50 (4): 10–16.

Keeley, P., and R. Harrington. 2010. *Uncovering student ideas in physical science, volume 1: 45 force and motion formative assessment probes.* Arlington, VA: NSTA Press.

Konicek-Moran, R. 2011. *Yet more everyday science mysteries.* Arlington, VA: NSTA Press.

NGSS Lead States. 2013. *Next Generation Science Standards: For states by states. www. nextgenscience.org/next-generation-science-standards.*

National Research Council (NRC). 2012. *A framework for K–12 science education: Practices, crosscutting concepts, and core ideas.* Washington, DC: National Academies Press.

INTERNET RESOURCE

NGSS Table: 3-PS2-2 Motion and Stability: Forces and Interactions
  www.nextgenscience.org/3-ps2-2-motion-and-stability-forces-and-interactions

NSTA CONNECTION

Download the "Swinging Pendulum" probe at *www.nsta.org/SC1310.*

● ●

Pendulums and Crooked Swings: Connecting Science and Engineering Reflection and Study Guide

QUESTIONS TO THINK ABOUT AFTER YOU READ THIS CHAPTER

1. What does the chapter quote, "the relationship between science and engineering practices is one of complementarity" mean to you (Bybee 2011)?

2. Science and engineering share several of the same practices. However, there are also differences. What are some of the differences between science and engineering?

3. How can an understanding of scientific concepts help students solve an engineering problem? Likewise, how can solving an engineering problem help students develop an understanding of scientific concepts?

4. How can the "Swinging Pendulum" probe be used to combine a core idea with a crosscutting concept and scientific practice?

5. What are some instructional advantages to using a probe like the "Swinging Pendulum" prior to having students plan and carry out a scientific investigation? How is this an example of how formative assessment informs next steps for instruction as well as promotes learning?

6. What are some of the formative data a teacher would collect to determine how well students can plan and carry out a scientific investigation? How could this data be used to provide feedback to students?

7. What is transfer of knowledge and why is it important to formatively assess learning transfer? How can engineering problems provide an ideal context for students to transfer their science learning to a new situation?

8. How is the Crooked Swing story an example of an engineering problem? How do students use their scientific knowledge to solve the problem? Which engineering practices do students use to solve the Crooked Swing problem?

9. How can you use formative assessment to determine how well students can use the engineering design process? How can you provide feedback to improve their ability to use the engineering design process?

10. How can reflective writing be used to help students reflect on the nature of science and engineering after completing a scientific investigation and engineering problem? How can this reflection process be used for feedback?

PUTTING FORMATIVE ASSESSMENT INTO PRACTICE

1. How does your curriculum or instructional materials link scientific investigations to engineering problems or use engineering problems to develop science concepts?

2. What did you learn about your students' readiness to undertake the Crooked Swing problem by examining their responses to the "Swinging Pendulum" probe? What instructional decisions did you make after eliciting students' ideas about pendulums?

3. How did you present the Crooked Swing story? Describe how you gathered evidence that students could use their pendulum knowledge to define the problem.

4. Did you use the checklist in Table 28.2 or some other formative assessment data collection method to observe students as they used the engineering design process? What was some of the feedback on the engineering design process you provided based on your observations?

5. Based on what you learned from using a formative assessment probe with your students prior to having them solve an engineering problem, what suggestions do you have for your colleagues and future teachers?

GOING FURTHER

1. Read and discuss the Teacher Notes for the "Swinging Pendulum" probe (Keeley and Harrington 2011 pp. 202–204). Pay particular attention to the Related Research and Suggestions for Instruction and Assessment sections.

2. Read and discuss the Teacher Notes for the Crooked Swing story (Konicek-Moran 2010).

3. Compare and contrast the science and engineering practices on the chart on pages 50–53 in *A Framework for K–12 Science Education* (NRC 2012) or online at *www.nap.edu/openbook.php?record_id=13165&page=53.*

4. Read and discuss the article: Bybee, R. 2011. Science and engineering practices in the K–12 classroom. *Science and Children* 48 (2): 10–16.

5. Read and discuss the guest editorial: Milano, M. 2013. The *NGSS* and engineering for young learners: Beyond bridges and egg drops. *Science and Children* 50 (2): 10–16.

6. Read and discuss the article: Crismond, D., L. Gellert, R. Cain, and S. Wright. 2013. Engineering encounters: Minding design missteps. *Science and Children* 51 (2): 80–85. This article provides a watch list of misconceptions for beginning designers that could be useful for formative assessment purposes.

7. Watch and discuss the NSTA-archived *NGSS* webinar on engineering practices in the *NGSS* at *http://learningcenter.nsta.org/products/symposia_seminars/NGSS/web-seminar15.aspx.*

Is It Melting? Formative Assessment for Teacher Learning

Formative assessment probes are effective tools for uncovering students' ideas about the various concepts they encounter when learning science. They are used to build a bridge from where the student is in his or her thinking to where he or she needs to be in order to construct and understand the scientific explanation for observed phenomena. In the process of using formative assessment, the function of teaching and learning shifts from one where children are merely provided with information about new concepts by the teacher to one where children have their own theories, which are challenged and reconstructed as new evidence is assimilated. To make this shift, teachers must first use formative assessment to find out what students' initial ideas are *prior* to instruction—as this is the conceptual framework the student will use to make sense of the new concepts and experiences they encounter in the science classroom. Second, this information needs to be used by the teacher to present new concepts or design experiences in a way that the student will be able to assimilate into their existing conceptual framework and recognize when their own ideas need to change.

While this constructivist view of teaching and learning works well with children, it also supports teacher learning. In this column, we will explore how the formative assessment probe "Is It Melting?" was used to confront teachers with their own alternative conceptions and help them reconstruct their ideas as they were presented with evidence that did not fit their initial conceptions. The following scenario is from a recent two-day formative assessment workshop for K–5 teachers conducted in a large school district. The afternoon session focused on using formative assessment probes to uncover students' ideas about physical changes.

Eliciting Teachers' Ideas

In grades K–2, the disciplinary core idea in Structure and Properties of Matter (PS1.A) states: "Different kinds of matter exist and many of them can be either solid or liquid, depending on temperature. Matter can be described and classified by its observable properties" (NGSS Lead States 2013). In grades 3–5, one of the disciplinary core ideas and performance expectations for Structure and Properties of Matter focuses on recognizing that matter is conserved during physical changes. These changes include heating or mixing substances. To make a bridge between the grades K–2 and grades 3–5 disciplinary core ideas, students need experiences identifying examples of and observing and explaining melting and dissolving in order to differentiate between these two physical changes. To help the teachers understand how their students might think about these two physical processes, I used the "Is It Melting" probe with the card sort strategy to engage them in discussing their ideas about physical changes, particularly melting and dissolving (Figure 29.1; Keeley, Eberle, and Farrin 2005). Each of the answer choices on the probe (A-I) were printed onto cards and the teachers sorted them into two groups: changes that are examples of melting and changes that are not examples of melting.

I assumed that the teachers would sort these appropriately and talk about what their students might think. Never make assumptions about student or teacher learning! I was surprised to find that most of the groups placed H, "sucking on a lollipop or hard candy" into the melting category. As I listened to their discussions, I heard a lot of talk about heat in the warm mouth melting the solid candy into a liquid. When we discussed the card sort as a large group, the teachers held fast to the notion that the lollipop or hard candy melted in our mouths. Several groups that initially placed it as a non-example of melting even changed their claim after listening to the arguments of others. A few groups defended the idea that it was dissolving rather than melting but lacked a cohesive explanation. Most claimed that because it was made of sugar, it dissolved.

At this point, I asked the group what we could do to test their claim that hard candies melt when we suck on them. One of the teachers came up with the clever idea that we could get some wrapped hard candies and suck on them with the wrappers on! If this was an example of melting because of the heat transferred from our mouth to the candy, then the solid candy would become liquid inside the wrapper. So, we decided to revisit this the next day. I would bring wrapped candies to our session, and we would test their claim.

Misconceptions: Not Just for Students

Fast forward to the next day. I passed out the wrapped candies and we sucked on them with the wrappers on. Lo and behold, there was no "melting" inside the wrapper! Typical of children's resistance to change, even when the evidence from observation is right there under your tongue so to speak, several of the teachers said we needed to suck on them longer and eventually they would melt. So we did, and still there was no change. Finally, the teachers were convinced that it was not an example of melting and they sided with the few groups that claimed it was an example of dissolving. We then unwrapped a candy and sucked on it and discussed the difference. One of the teachers even looked up the melting point of sugar on her smart phone and found that it is much higher than the temperature inside our mouths!

FIGURE 29.1.

The Is It Melting? Probe

Is It Melting?

The list below involves situations that cause changes in materials. The materials are *italicized*. Put an X next to the situations in which the *italicized* materials undergo melting.

_____ **A** Putting a bowl of frozen *ice cream* in the sun.

_____ **B** Sawing *wood* to make sawdust.

_____ **C** Dissolving *salt* in water.

_____ **D** Adding a *LifeSaver* candy to a glass of warm water.

_____ **E** *Water* evaporating from a pan.

_____ **F** Dissolving a *sugar cube* in a cup of hot tea.

_____ **G** Pouring vinegar on *baking soda*.

_____ **H** Sucking on a *lollipop* or other *hard candy*.

_____ **I** Holding an *ice cube* in your hand.

Explain your thinking. Describe the "rule" or reasoning you used to decide if something melts.

The teachers now had a partially correct explanation—that the sugar in the candy dissolved in the liquid saliva in the mouth, hastened by the warm temperature. But they needed more information to fit into this conceptual framework to deepen their understanding. But rather than filling their heads with information delivered by the teacher to the student, I decided to use the "Scientist Idea Comparison" strategy to activate their thinking and provide them with new information to consider.

Using the explanation in the teacher notes for the probe (Keeley, Eberle, and Farrin 2005), I shared the following on a PowerPoint slide, saying here is how a scientist might explain the probe: "There are two examples of melting, *A* and *I*. Melting is a process in which a solid undergoes changes in the arrangement and average motion of particles to become a liquid. In order to melt, a substance needs to absorb heat energy (thermal energy). This thermal energy increases the average motion of the particles, resulting in a change in state from solid to liquid. Dissolving is not a change in state. When solid materials such as the hard candy dissolve, they involve intermolecular forces that break down the substance (hard candy) into smaller particles (molecules). The interaction between the molecules of sugar and the molecules of water is stronger than the interaction between the sugar molecules. We can think of these two different processes as interactions—Dissolving is an interaction between two materials: a solute (the candy) and the solvent (water or, in this case, saliva). Melting involves only one material which gains heat energy" (Keeley, Eberle, and Farrin 2005, p. 74). The explanations for the probes in the Teacher Notes are simplified explanations, with just enough for the teacher to understand the concept. Additional resources can be used to build deeper understanding.

Using the Scientists' Idea Comparison strategy, the teachers discussed which of their ideas were similar to the scientists' and which were different. This metacognitive strategy helped them assimilate this new information into their new framework and at the same time revealed whether there were still any lingering ideas they were struggling with that needed to be addressed.

The Power of Formative Assessment

Most important, this professional development episode modeled for the teachers what can happen in the science classroom. By experiencing a tenaciously held incorrect idea themselves and working through it so that they could eventually give up their prior ideas in favor of a better explanation, they understood firsthand the power of using formative assessment to drive instruction and promote conceptual learning. They also realized how using formative assessment probes and the teacher notes that accompany them can sometimes surface their own misunderstandings and help them work through the process of developing scientific understanding. Furthermore, the teachers realized how powerful the learning was for them as they interacted in small and large groups. Whether you provide professional development for teachers, lead professional learning communities, mentor other teachers, or teach preservice teachers, formative assessment probes are effective resources for uncovering and developing teachers' ideas related to the disciplinary core ideas and performance expectations in the *Next Generation Science Standards*.

REFERENCES

Keeley, P., F. Eberle, and L. Farrin. 2005. *Uncovering student ideas in science, volume 1: 25 formative assessment probes.* Arlington, VA: NSTA Press.

NGSS Lead States. 2013. *Next Generation Science Standards: For states, by states.* Washington, DC: National Academies Press. *www.nextgenscience.org/next-generation-science-standards.*

NSTA CONNECTION

Download the "Is It Melting?" probe at *www.nsta.org/SC1311*.

• •

Is It Melting?
Formative Assessment for Teacher Learning
Reflection and Study Guide

QUESTIONS TO THINK ABOUT AFTER YOU READ THIS CHAPTER

1. How can you use a bridge analogy to describe the use of formative assessment?

2. Describe the instructional shift teachers must make in order to use formative assessment effectively.

3. What is constructivism? How does the use of formative assessment probes mirror the constructivist view of teaching and learning?

4. How can the *Uncovering Student Ideas in Science* probes be used to support teacher learning? Describe an example of a probe you used that helped you learn more about the content and pedagogy of the science you teach.

5. How would you have answered the "Is It Melting?" probe prior to reading this chapter? What did you learn about melting and dissolving as a result of reading this chapter?

6. How did the facilitator's and teachers' experiences in the workshop described in the chapter mirror what happens in the elementary classroom?

7. Why did the facilitator use the "Scientist's Idea Comparison" strategy before explaining the science to the teachers? How would you use this technique with students?

8. How did the teachers' experience in the workshop help them better understand the benefit of using formative assessment to inform teaching and promote learning in their classroom?

PUTTING FORMATIVE ASSESSMENT INTO PRACTICE

1. How does your curriculum address the physical changes of melting and dissolving? How do you help students distinguish between these changes at your grade level?

2. What did you learn about your students' ideas by examining their responses to the probe? Were you surprised by any of their responses?

3. Did you combine the probe with any formative assessment classroom techniques (FACTs) such as a card sort? What instructional decisions did you make as a result of using this probe?

4. This chapter was primarily focused on using formative assessment probes for teacher learning. What did you learn as a result of using this probe?

5. Formative assessment probes can be used in a variety of teacher learning formats ranging from individual learning to collaborative group learning in a variety of structures. What ideas do you have for using formative assessment probes for teacher learning with your colleagues?

6. Based on what you learned from using this probe with your students (or with teachers), what suggestions do you have for your colleagues (including teacher education and professional development) and future teachers?

GOING FURTHER

1. Read and discuss the Teacher Notes for the "Is It Melting?" probe (Keeley, Eberle, and Farrin 2005, pp. 74–77). Pay particular attention to the Related Research and Suggestions for Instruction and Assessment sections.

2. Read and discuss the introduction chapter in *Uncovering Student Ideas in Science, Vol. 3* (Keeley, Eberle, and Dorsey 2008), which focuses on using formative assessment probes for teacher learning. This chapter is available as a free download through NSTA Press at *www.nsta.org/store/product_detail.aspx?id=10.2505/9781933531243*.

3. Examine different formative assessment classroom techniques (FACTs) that can be used with students or with teachers. When FACTs are combined between the three books in the FACTs series (Keeley 2008, 2014; Keeley and Tobey 2011), there are 136 different FACTs that can be used with students and with teachers.

Chapter 30

Is It Made of Parts? Scaffolding a Formative Assessment Probe

All living things, from a tiny, single-celled bacterium to an enormous blue whale, are made of parts. These parts have specific functions that help organisms carry out their life processes. Parts are made up of even smaller parts—all organisms are made up of cells, which contain smaller parts within the cell, which are made up of molecules. Some parts are combined into systems that are specialized to carry out a particular function. For example, cells in multicellular organisms combine to form tissues that make up organs that carry out a specific function. Parts and wholes and their functions is a recurring concept in life science that begins with young children learning about external parts of familiar organisms and builds through successive grade levels culminating in understanding the parts of cells that carry out chemical reactions or contain genetic information, and the biomolecules involved.

One of the disciplinary core ideas in *A Framework for K–12 Science Education* is LS1.A Structure and Function (NRC 2012). This disciplinary core idea is included in the *Next Generation Science Standards (NGSS)*, which state that grade 1 students are expected to use the idea that "all organisms have external parts. Different animals use their body parts in different ways to see; hear; grasp objects; protect themselves; move from place to place; and seek, find, and take in food, water, and air. Plants also have different parts (roots, stems, leaves, flowers, fruits) that help them survive and grow" (NGSS Lead States 2013, p. 12). The formative assessment probe "Is It Made of Parts?" (Figure 30.1, p. 211) is designed to uncover K–2 children's initial ideas about the parts of organisms (Keeley 2013). The teacher can then use this assessment information to make informed decisions about scaffolding the instructional opportunities children will need to develop a foundational understanding of the relationship between structure and function.

Using the Probe

To scaffold this "structure and function" probe for assessment purposes, begin by identifying the sub-ideas that this formative assessment probe can uncover at the K–2 level. These sub-ideas for formative assessment include:

1. Recognizing that all organisms have parts (with a focus on the external structures, not internal structures),

2. Describing how animals use their body parts,

3. Identifying parts of a plant, and

4. Describing basic functions of plant parts (to survive and grow).

The probe is purposefully designed to uncover these sub-ideas using the organisms and parts of organisms in the pictures. Starting with sub-idea 1, notice that the language of the probe uses "parts," not structures. Later, children will learn that parts of living things are referred to as structures, but first it is best to start with the familiar terminology. Also notice that the examples are plants and animals or parts of plants and animals in order to develop the generalization that all organisms have parts. In some of the examples, the parts are obvious; in others, they are not obvious.

Begin by having children choose the things on the list that are made of parts and describe the "rule" they used to decide if they are made of parts. Some of the typical responses from primary age children include: if there are arms or legs; if there are eyes, ears, noses, or mouths; and if it doesn't all look the same. Probing further you might find that students fail to include the worm as it doesn't have the familiar body parts. Or they may fail to recognize that the snake has parts if their rule is that it has to have arms or legs, even though it has a head, eyes, and so on. Some may not include the leaf or feather because it is a part of something else and the seed looks the same all around. All of this information can be used to design a lesson on external "parts and wholes" to show how plants and animals have a variety of different parts and their parts are also made up of smaller parts. If students are still stuck on the earthworm and the seed, have them observe real earthworms and seeds, using magnifying devices, to see different parts. They might see that a seed has an outer coat or a part where the seed attached to the plant. They might notice the segments on an earthworm, the light-colored band (clitellum), and sometimes they may even see the bristles on a large earthworm or the mouth opening if they use a magnifying lens. This might be a time to find a children's book about earthworms that shows the external structure of an earthworm. Listen carefully for evidence of how their initial ideas about parts of organisms are changing as they are

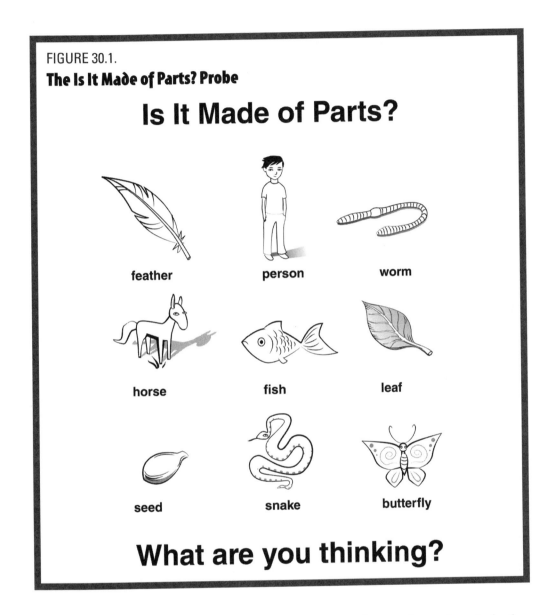

FIGURE 30.1.
The Is It Made of Parts? Probe

Is It Made of Parts?

feather   person   worm

horse   fish   leaf

seed   snake   butterfly

What are you thinking?

confronted with their ideas. Use additional organisms and parts of organisms to further challenge, refine, and solidify their thinking. This information will help you determine the extent to which students can use sub-idea 1.

After students recognize that all of the organisms listed on the probe have parts, and that some of the examples are parts of organisms, it is time to assess sub-idea 2: how they connect parts (structure) to uses (function), beginning with animals. Ask them which things on the list are animals or parts of animals (note that some children may not consider a person, worm, butterfly, or snake to be an animal if their concept of an animal is limited to mammals, which indicates the need to develop the precursor idea of "animal"). Starting with the whole organisms, have the children point out the different parts of the animals on the list (person, horse, fish, snake, butterfly, and worm) and ask them what they think the animal uses that part for. Start with a familiar animal like the person or the horse and then move to the other animals on the list. They might not know what the parts of an earthworm are called or what they do, but give them an opportunity to share what they think the parts do for the worm. Probe further by asking them which parts of the animals help the animal see; hear; grasp objects; protect themselves; move from place to place; and seek, find, and take in food, water, and air. As their ideas are developing, add examples of animals to probe further and examine their thinking about functions of animal parts. Then move to the feather and ask them if that is an animal or a part of an animal. If students recognize that a feather is a part of a bird, ask them what the bird uses its feathers for. Probe further to see if they recognize that a part of an animal (a feather) is made of parts. Have a feather available for students to observe. For example, students might notice the many barbs of the feather that join together to make up the blade of the feather.

Now that you have information about students' ideas related to sub-idea 2—the uses (function) of animal parts (structures)—and have informed your instruction to develop and solidify these ideas, you can move on to sub-idea 3. Show students an actual plant or picture of a plant. Ask them, "If animals have parts, do plants also have parts?" Provide them with an opportunity to point out and name the plant parts they can see. Ask them which examples in the pictures are parts of plants (leaf and seed). Have them name other parts of plants that aren't in the pictures, guiding them toward roots, flower, fruit, and stem if they have not already mentioned these. The information reveals which parts they are familiar with and which parts they may need more opportunities to observe and discuss.

The last sub-idea has students connect the plant parts to their functions, without going into detailed plant processes at this level, such as photosynthesis. Ask them what they think the leaf and seed are used for. Listen carefully to see if they recognize that plants need sunlight to grow and the leaf helps the plant "get" sunlight. Listen for early ideas about reproduction, such as seeds are used to grow more plants. Extend the probe to include other plant parts in this sub-idea such as roots, stems, flowers, and fruits.

The "Is It Made of Parts" probe is an example of scaffolded formative assessment. Using a formative assessment probe to elicit each of the sub-ideas that make up a learning goal provides valuable information that a teacher can then use to design instruction that builds upon and links sub-ideas, while simultaneously promoting student thinking. As you use any of the formative assessment probes in the *Uncovering Student Ideas in Science* series, think about how you can break any learning goal down into a smaller set of ideas that you can use to scaffold assessment and instruction.

REFERENCES

Keeley, P. 2013. *Uncovering student ideas in primary science: 25 new formative assessment probes.* Arlington, VA: NSTA Press.

National Research Council (NRC). 2012. *A framework for K–12 science education: Practices, crosscutting concepts, and core ideas.* Washington, DC: National Academies Press.

NGSS Lead States. 2013. *Next Generation Science Standards: For states, by states.* Washington, DC: National Academies Press. *www.nextgenscience.org/next-generation-science-standards.*

NSTA CONNECTION

Download the "Is It Made of Parts?" probe at *www.nsta.org/SC1312.*

• •

Is It Made of Parts?
Scaffolding a Formative Assessment Probe
Reflection and Study Guide

QUESTIONS TO THINK ABOUT AFTER YOU READ THIS CHAPTER

1. How does the concept of "parts and wholes" contribute to students' understanding of the core idea of structure and function?

2. How does the probe "Is It Made of Parts?" inform teachers about students' initial ideas related to the parts of organisms?

3. What is an instructional scaffold? When have you used scaffolding in your teaching? How is scaffolding used with formative assessment?

4. What are sub-ideas? Why is it useful to identify sub-ideas for formative assessment? What should one look for when examining sub-ideas for elementary students?

5. How does the formative assessment probe, "Is It Made of Parts?" address sub-ideas?

6. What are some of the ideas students have about parts of organisms? Why is it helpful to be aware of these common ideas before using the probe with your students?

7. How did this chapter describe ways to scaffold this probe? What are some examples of instructional decisions that were made through the use of scaffolding?

8. How did each sub-idea link to the next sub-idea and what role did formative assessment play in linking together sub-ideas?

PUTTING FORMATIVE ASSESSMENT INTO PRACTICE

1. When and how does your curriculum or instructional materials address parts of organisms?

2. What did you learn about your students' ideas by examining their responses to the probe? Were you surprised by any of their responses?

3. Did you scaffold this probe as described in the chapter? How did breaking down into sub-ideas in order to scaffold the probe enhance the process of formative assessment?

4. As a result of reading this chapter, how might you use scaffolding with other formative assessment probes?

5. Based on what you learned from using this probe with your students, what suggestions do you have for your colleagues and future teachers?

GOING FURTHER

1. Read and discuss the Teacher Notes for the "Is It Made of Parts?" probe (Keeley 2013, pp. 22–24). Pay particular attention to the Related Research and Suggestions for Instruction and Assessment sections. This probe is available as a free download through NSTA Press at *www.nsta.org/store/product_detail. aspx?id=10.2505/9781936959518*.

2. Parts of organisms is addressed in the *NGSS* and *A Framework for K–12 Science Education* under Core Idea LS1.a: Structure and Function. Read the sections on pages 143–145 or online at *www.nap.edu/openbook.php?record_id=13165&page=143*.

3. Project 2061's *Resources for Science Literacy* provides an example of how lesson design principles were used to scaffold an "Exploring Parts and Wholes' lesson for young children. You can access this lesson online at *www.project2061.org/ publications/rsl/online/GUIDE/CH2/OPARTS.HTM*.

References

Ainsworth, L., and D. Viegut. 2006. *Common formative assessments.* Thousand Oaks, CA: Corwin Press.

Ansberry, K., and E. Morgan. 2009. Secrets of seeds. *Science and Children* 46 (5): 161–168.

Ashbrook, P. 2012. Shining light on misconceptions. *Science and Children* 49 (2): 30–31.

Barman, C., N. Barman, K. Berglund, and M. Goldston. 1999. Assessing students' ideas about animals. *Science and Children* 37 (1): 44–49.

Barman, C., N. Barman, M. Cox, K. Berglund-Newhouse, and M. J. Goldston. 2000. Students' ideas about animals: Results of a national study. *Science and Children* 38 (1): 42–47.

Barrows, L. 2007. Bringing light onto shadows. *Science and Children* 44 (9): 43–45.

Baxter, L., and M. Kurtz. 2001. When a hypothesis is not an educated guess. *Science and Children* 38 (8): 18–20.

Brkich, K. 2012. Science shorts: Is concrete a rock? *Science and Children* 50 (4): 80–82.

Bybee, R. 2011. Science and engineering practices in the K–12 classroom. *Science and Children* 48 (2): 10–16.

Concannon, J., P. Brown, and E. Pareja. 2007. Making the connection: Addressing students' misconceptions of circuits. *Science Scope* 30 (3): 10–15.

Crismond, D., L. Gellert, R. Cain, and S. Wright. 2013. Engineering encounters: Minding design missteps. *Science and Children* 51 (2): 80–85.

Fitzsimmons, P., D. Leddy, L. Johnson, S. Biggam, and S. Locke. 2013. The moon challenge. *Science and Children* 50 (1): 36–41.

Gagnon, M., and S. Abell. 2008. Perspectives. *Science and Children* 45 (4): 60–61.

Harvard-Smithsonian Center for Astrophysics. 1995. *Private universe project in science.* Washington, DC: Harvard-Smithsonian Center for Astrophysics.

Haysom, J., and M. Bowen. 2010. *Predict, observe, explain: Activities enhancing scientific understanding.* Arlington, VA: NSTA Press.

Hodgson-Drysdale, T., and E. Ballard. 2011. Explaining electrical circuits. *Science and Children* 48 (8): 37–41.

Keeley, P. 2008. *Science formative assessment: 75 practical strategies for linking assessment, instruction, and learning.* Thousand Oaks, CA: Corwin Press.

Keeley, P. 2013. *Uncovering student ideas in primary science.* Arlington, VA: NSTA Press.

Keeley, P. 2014. *Science formative assessment: 50 more practical strategies for linking science, instruction, and assessment.* Thousand Oaks, CA: Corwin Press.

Keeley, P., F. Eberle, and C. Dorsey. 2008. *Uncovering student ideas in science, volume 3: Another 25 formative assessment probes*. Arlington, VA: NSTA Press.

Keeley, P., F. Eberle, and L. Farrin. 2005. *Uncovering student ideas in science, volume 1: 25 formative assessment probes*. Arlington, VA: NSTA Press.

Keeley, P., F. Eberle, and J. Tugel. 2007. *Uncovering student ideas in science, volume 2: 25 more formative assessment probes*. Arlington, VA: NSTA Press.

Keeley, P., and R. Harrington. 2010. *Uncovering student ideas in physical science, volume 1: 45 new force and motion assessment probes*. Arlington, VA: NSTA Press.

Keeley, P., and R. Harrington. 2014. *Uncovering student ideas in physical science, volume 2: 39 new electricity and magnetism formative assessment probes*. Arlington, VA: NSTA Press.

Keeley, P., and C. Sneider. 2012. *Uncovering student ideas in astronomy: 45 new formative assessment probes*. Arlington, VA: NSTA Press.

Keeley, P., and C. Tobey. 2011. *Mathematics formative assessment: 75 practical strategies for linking assessment, instruction, and learning*. Thousand Oaks, CA: Corwin Press.

Kenyon, L., C. Schwartz, and B. Hug. 2008. The benefits of scientific modeling. *Science and Children* 45 (2): 40–44.

Koba, S. 2011. *Hard to teach science concepts: A framework to support learners, grades 3–5*. Arlington, VA: NSTA Press.

Konicek-Moran, R. 2008. *Everyday science mysteries*. Arlington, VA: NSTA Press.

Konicek-Moran, R. 2010. *Even more everyday science mysteries*. Arlington, VA: NSTA Press.

Legaspi, B., and W. Straits. 2011. Living or Nonliving? First grade lessons on life science and classification address misconceptions. *Science and Children* 48 (8): 27–31.

Matkins, J., and J. McDonnough. 2004. Circus of light. *Science and Children* 41 (5): 50–54.

McComas, W. 1996. Ten myths of science: Reexamining what we think we know. *School Science & Mathematics* 96: 10.

Michaels, S., A. W. Shouse, and H. A. Schweingruber. 2008. *Ready, set, science! Putting research to work in K–8 science classrooms*. Washington, DC: National Academies Press.

Milano, M. 2013. The *NGSS* and engineering for young learners: Beyond bridges and egg drops. *Science and Children* 50 (2): 10–16.

Mundry, S., P. Keeley, and C. Landel. 2010. *A leader's guide to science curriculum topic study*. Thousand Oaks, CA: Corwin Press.

National Research Council (NRC). 2012. *A framework for K–12 science education: Practices, Crosscutting Concepts, and Core Ideas*. Washington, DC: National Academies Press.

Nelson, G. 2008. Building ladders to the stars. *Science and Children* 45 (1): 8.

NGSS Lead States. 2013. *Next Generation Science Standards: For states, by states*. Washington, DC: National Academies Press. *www.nextgenscience.org/next-generation-science-standards*.

Palmieri, A., A. Cole, S. deLisle, E. Erickson, and J. Janes. 2008. What's the matter with teaching children about matter? *Science and Children* 46 (4): 20–23.

Pea, C., and D. Sterling. 2002. Cold facts about viruses. *Science Scope* 25 (3): 12–17.

Pine, J. 1999. To hypothesize or not to hypothesize. In *Foundations: Inquiry*, Institute for Inquiry, 61. Washington, DC: National Science Foundation.

Roberts, D., C. Bove, and E. van Zee. 2007. *Teacher research: Stories of learning and growing*. Arlington, VA: NSTA Press.

Ross, D., D. Fisher, and N. Frey. 2009. The art of argumentation. *Science and Children* 47 (3): 28–31.

Rutherford, F. J., and A. Ahlgren. 1988. *Science for all Americans*. Washington, DC: AAAS.

Schussler, E., and J. Winslow. 2007. Drawing on students' knowledge. *Science and Children* 44 (4): 40–44.

Schwartz, R. 2007. What's in a word? How word choice can develop (mis)conceptions about nature of science. *Science Scope* 31 (2): 42–47.

Smith, L., D. Sterling, and P. Moyer-Packenham. 2006. Activities that really measure up. *Science and Children* 44 (2): 30–33.

Smithenry, D., and J. Kim. 2010. Beyond predictions. *Science and Children* 48 (2): 48–52.

Stavy, R., and D. Tirosch. 2000. *How students (mis)understand science and mathematics: Intuitive rules*. New York: Teachers College Press.

Stavy, R., and N. Wax. 1989. Children's conceptions of plants as living things. *Human Development* 32: 88–94.

Tugel, J., and I. Porter. 2010. Uncovering student thinking in science through CTS action research. *Science Scope* 33 (1): 30–36.

Tweed, A. 2009. *Designing effective instruction: What works in science classrooms*. Arlington, VA: NSTA Press.

Volkman, M., and S. Abell. 2003. Seamless assessment. *Science and Children* 40 (8): 41–45.

Watson, B., and R. Konicek. 1990. Teaching for conceptual change: Confronting children's experience. *Phi Delta Kappan* 680–684.

Index